26th SNAME Offshore Symposium 2021

AA005484

Held online

Houston, Texas, USA
6-7 April 2021

ISBN: 978-1-7138-3136-5

Printed from e-media with permission by:

Curran Associates, Inc.
57 Morehouse Lane
Red Hook, NY 12571

Some format issues inherent in the e-media version may also appear in this print version.

Printed with permission by Curran Associates, Inc. (2021)

For permission requests, please contact Society of Naval Architects and Marine Engineers (SNAME) Texas Section at the address below.

Society of Naval Architects and Marine Engineers (SNAME) Texas Section
99 Canal Center Plaza, Suite 310
Alexandria, VA 22314
USA

Phone: +1-703-997-6701

www.sname.org/texassection

Additional copies of this publication are available from:

Curran Associates, Inc.
57 Morehouse Lane
Red Hook, NY 12571 USA
Phone: 845-758-0400
Fax: 845-758-2633
Email: curran@proceedings.com
Web: www.proceedings.com

TABLE OF CONTENTS

A NOVEL DATA-DRIVEN SUBSEA PROJECT DELIVERY ECOSYSTEM

Gautier Noiray
McDermott

Matt Boisne
McDermott

Proceedings of the 26th Offshore Symposium, April 6th an 7th 2021, Houston, Texas
Texas Section of the Society of Naval Architects and Marine Engineers

ABSTRACT

Beyond the craze for Digitization in the industry, there remains a lack of clarity on the concrete changes and value most companies are extracting from it. The recent market conditions, exacerbated by the current global pandemic, make this question more relevant than ever, as any new investment becomes ever more challenged.

McDermott have been at the forefront of the energy industry digitization, and over the past couple of years, they have ideated, developed, and implemented a novel data-driven Subsea Project delivery ecosystem. This cross-functional ecosystem leverages cloud technology and Python scripting to enhance and systemize processes, fundamentally tying back to the field architecture digital twin. The purpose of the paper is to communicate the main drivers at the origin of the initiative, to describe the current solution, including the associated risks (and mitigations) and the benefits that arise from its use as well as give the reader a sense of the more long-term vision.

Keywords: Subsea Field Development, Subsea Engineering, Digital Twin, Automation, Digitization, Costing

Introduction

Although the oil and gas industry has made great strides during the 20th century toward a more abundant and affordable energy, it has yet to take full advantage of the latest developments on the digital front [Ref. 1]. Digital technology has reshaped many industries such as the transport industry, mass media, and retail industry, just to name a few. However, most of the day-to-day activities in the world of subsea field developments do not seem to have capitalized on the last two decades of technological progress.

The purpose of this paper is to discuss a new working model which achieves a greater level of efficiency and collaboration for the conception and execution of subsea projects. And, although the focus will be around subsea developments, the framework discussed is very relevant to other business lines that involve multiple disciplines and interfaces. The paper first lists the struggles and pain points that gave this initiative a direction. Then, it describes the solution selected to alleviate the aforementioned struggles. After that, it provides concrete examples, discusses their benefits and applications. Finally, it concludes with some forward-looking statements.

Motivations and Challenges

As an Engineering, Procurement, Construction, and Installation (EPCI) company, McDermott designs and delivers complex projects. The three most important success criteria are:

1. Safety of the people and the environment
2. On-time delivery: execute project per agreed-upon schedule
3. On-budget execution: execute project within budget

The first point is paramount and, while opportunities to integrate it to the ecosystem have been identified, more work is planned by the end of the year to bring the concept to a usable solution. As such, it will not be covered in this overview.

The last two points, although mentioned separately, are actually closely related. It is unfortunately quite common to see major projects in the construction industry being delivered late and above budget. While the consequences of schedule and budget overruns could have lasting effects a decade ago, those nowadays have a direct impact on a company's viability. As such, schedule and cost certainty are the top biggest motivators of this initiative.

Historically, projects are carried out by going through various phases. Typically, operators perform internal engineering to ensure the overall economics of a project. Then, an engineering or EPCI firm is consulted to further screen, optimize and engineer a solution. Finally, an EPCI firm is contracted to perform detailed engineering, procurement, construction and installation.

Going through these various project phases involves several hand-overs of information, requirements and asks. This inherently creates risks of value erosion by:

- Losing decision rationale (example: Unnecessary high volume of clarifications and qualifications)
- Time and cost impact of numerous input data querying
- Engineering rework
- Costly and time-consuming data handover.

The execution of a project involves several disciplines and stakeholders. Each discipline must go through a set of processes and tools.

Traditionally, those processes are not only carried out sequentially but also in silos between and within functions. Similarly, the flow of information follows a convoluted and inefficient pattern. Thus, resulting in the delivery of the final product at the very end of the endeavor with limited time for a meaningful optimization. Additionally, changes made by a function in their scope may not be communicated early enough and therefore impact negatively other functions by omission. This can have very dire consequences on budget and schedule. Those challenges encompass not only engineering and marine operations but also estimating, procurement and planning.

For the sake of expediency, work performed on previous projects is often used as a starting point for new ones. While time may be saved upfront, such practice will impede in delivering a bespoke solution.

SOLUTION

In order to efficiently tackle the problematics discussed above, it became apparent to McDermott there was a need for an open, user-friendly, scale-able platform as a foundation. There was no appetite to develop in-house the solution from scratch because a software company would propose a better solution within a much shorter timeframe. Within the 2016 market, FieldAp, which would then later grow into FieldTwin, presented itself as the best balance between product maturity, ease of integrating the user's need, strength of the developing team and the overall vision and strategy of FutureOn. FieldTwin allows the teams to quickly build a field architecture from a generic list of assets and thus, obtain a dimensionally and geographically correct representation of the field. In addition to that, each asset having the flexibility to be defined by several attributes, the field architecture also becomes a digital repository to the basis of design. Team members are invited with view/access rights reflecting their role in the project. With FieldTwin in its core, Subsea^{XD} would then be developed using McDermott own resources and EPCI know-how. From there, teams can trigger processes, iterate and evaluate a variety of solutions in an accelerated way that brings further value to the model by integrating system functional properties beyond simple geometric information.

Figure 1: FieldTwin viewer with asset metadata

FutureOn's FieldTwin platform leverages a RESTful API (Representational State Transfer Application Programming Interface) technology that allows users to perform actions using lines of code.

For instance, a flowline can be characterized by:

- Length
- Start coordinates
- End coordinates
- Depth profile
- Route (X,Y)
- Outer diameter
- Wall Thickness
- ...

The RESTful API allows a programmer to write a simple script that will be able to fetch all the above parameters of a flowline, using its unique URL (Uniform Resource Locator). Those same lines of code can be used to fetch the same information of a different flowline, by changing the URL. This is at the heart of the initiative.

The digital twin of the field architecture becomes the collaborative portal that project team members access to contribute to the project. Figure 1 shows the main layout of FieldTwin while Figure 2 illustrates the flow between the FutureOn FieldTwin platform and the Subsea^{XD} modules and how they communicate through APIs and scripts.

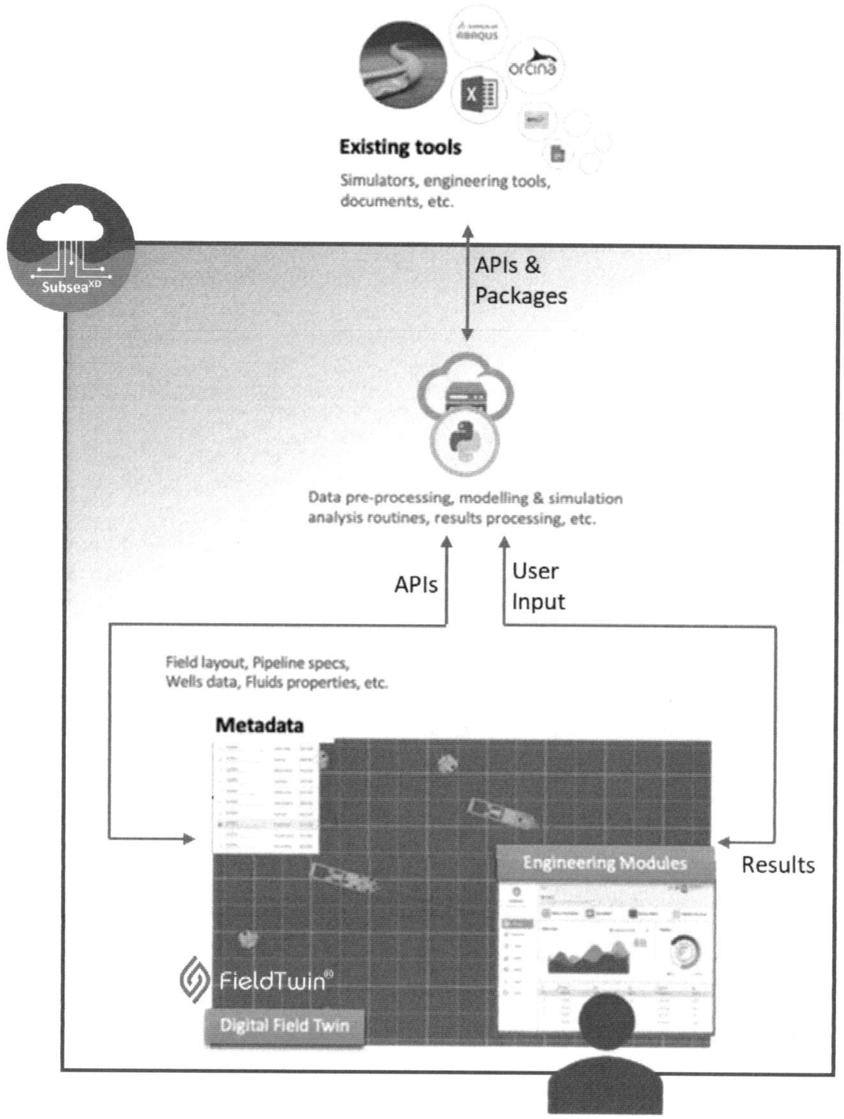

Figure 2: Modules data flow

In most instances, modules are hosted in Tabs inside of FieldTwin. They communicate with the FieldTwin project through Event Listeners that, when triggered, collect project data through REST APIs and send that information to various virtual machines.

At that point, a new load case is created automatically, which consists of a set of python scripts in which the right inputs were injected, based on the data received from FieldTwin. The scripts run on the virtual machine and output a number of data frames that are passed back to the front end of the application and formatted as required. In addition to data frames, the outputs can also be images, graphs, software files (Excel, Orcaflex, PipeSim, Abaqus…).

This architecture allows processes to be tag-centric and therefore, ensure a proper traceability of data and the various changes that may occur during a project.

Developing such solution is a very ambitious undertaking that requires a strong multifunctional commitment as well as a rigorous application of the company's procedures, including a very stringent scrutiny on security since internal and external confidential and proprietary data is handled.

Building such an architecture requires the combination of several skill sets. In fact, McDermott has built an agile team of engineers, with programming acumen, that works hand in hand with a web development team. Doing so ensures that best practices are adopted on all fronts. In addition to this core team, all functions provide resources on an ad hoc basis at all stages of the development, from proof of concept to final testing.

As the scope of use of such data-centric ecosystem expands, the volume of generated and processed data is increasing exponentially. It is key to anticipate early on the challenges associated with such expansion by including people with strong big data knowledge in the choices of architecture, software, processes, and infrastructure.

To provide an estimate of the volume of resources required, Figure 3 shows a breakdown of effort spent supporting the development of the Subsea^{XD} eco-system in 2020.

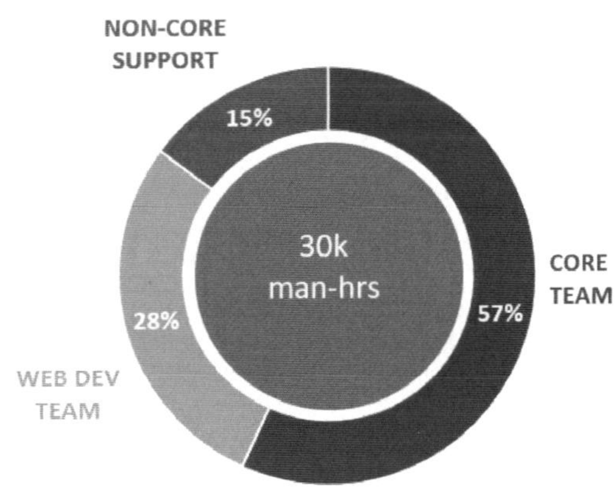

Figure 3: Summary of development effort breakdown for 2020

Although the initiative started with a focus on engineering processes it quickly became clear that the value addition transcends that function. In fact, most functions involved on a project rely heavily on excel tools and make use of mostly the same data. Therefore, expanding the Subsea^{XD} scope to other functions naturally contributes to, not only consolidating those function's processes, but also enhancing the collaboration across teams. After engineering, the next function that was incorporated into the ecosystem was marine operations, which naturally paved the way for procurement, planning and estimation. Figure 4 shows the ecosystem multi-functional characteristics and its main benefits.

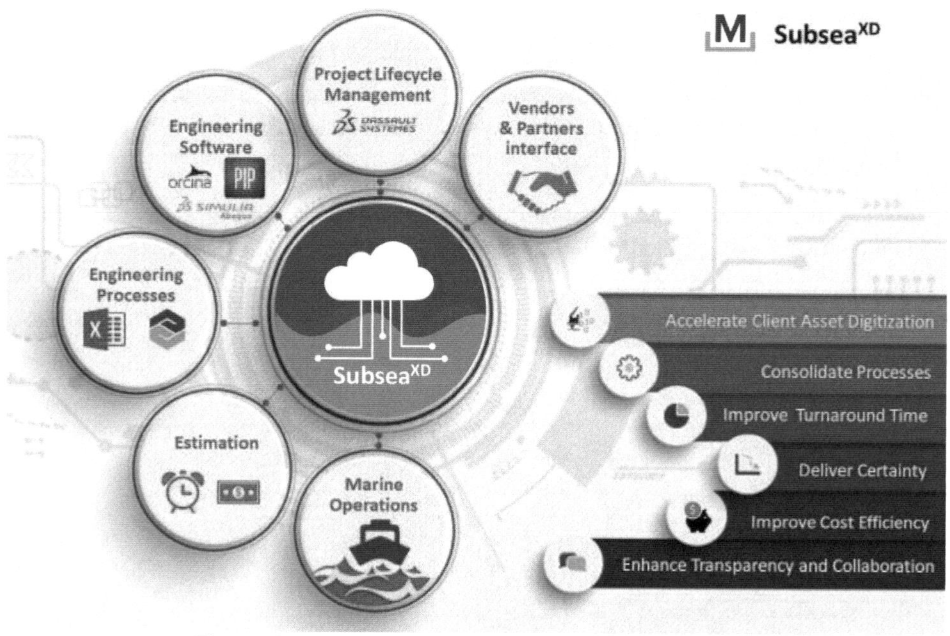

Figure 4: An ecosystem enabled by a digital design twin

Beyond enhancing internal collaboration, the ecosystem allows integration of third-party entities such as vendors and clients, provided those are favorable to this digital way of working.

From its inception, Subsea^{XD} has been developed to be fully integrated with McDermott's digital ecosystem. As such, it syncs up with McDermott's Project Lifecycle Management (PLM) platform to enable a tag-centric project controls approach. The field architecture becomes the gateway into project data such as reports, schedule, and man-hours tracking. Beyond the obvious internal benefits (see Figure 5 for three examples) that arise from the adoption of this ecosystem, another advantage is the opportunity for customers to maintain a digital stream that flows from them to McDermott and back, as the project goes through its life cycle. This un-interrupted flow accelerates the creation of an operational twin that will live on during the life of the asset.

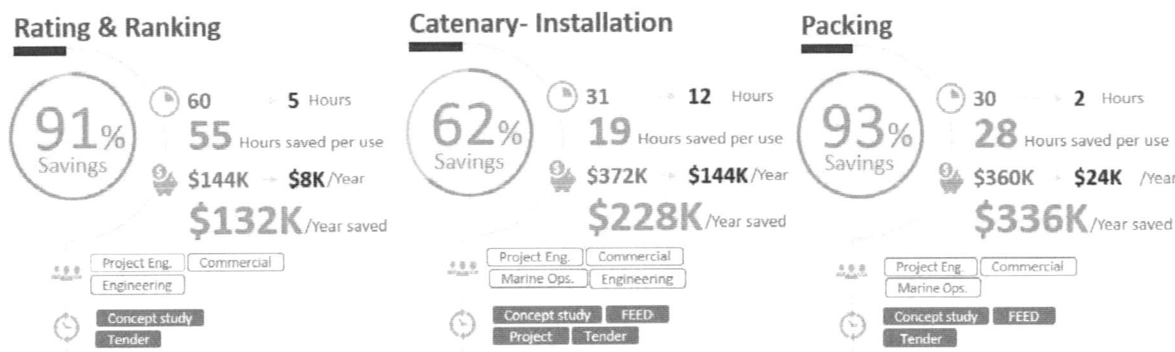

Figure 5: Internal savings achieved

As the Subsea^{XD} ecosystem grows, it incorporates more modules following the Agile development methodologies. Currently, 14 modules are at TRL 9 (Technology Readiness Level) as shown on Figure 6. It is worth noting that not all modules require the same effort of development. The majority of the TRL9 modules were prioritized due to the added value they would bring once in use.

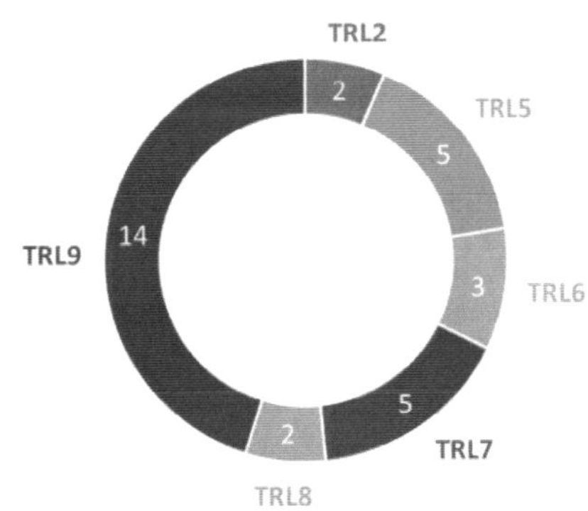

Figure 6: Current TRL distribution of Subsea^{XD} modules

Examples

Figure 7 illustrates, at a high level, the different blocks and flow of data, that come into play to ultimately generate the cost and schedule of a subsea project. At first glance, it is clear that three distinct blocks come into play and that data flows left to right, starting from basic field architecture requirements and ending as a cost and schedule.

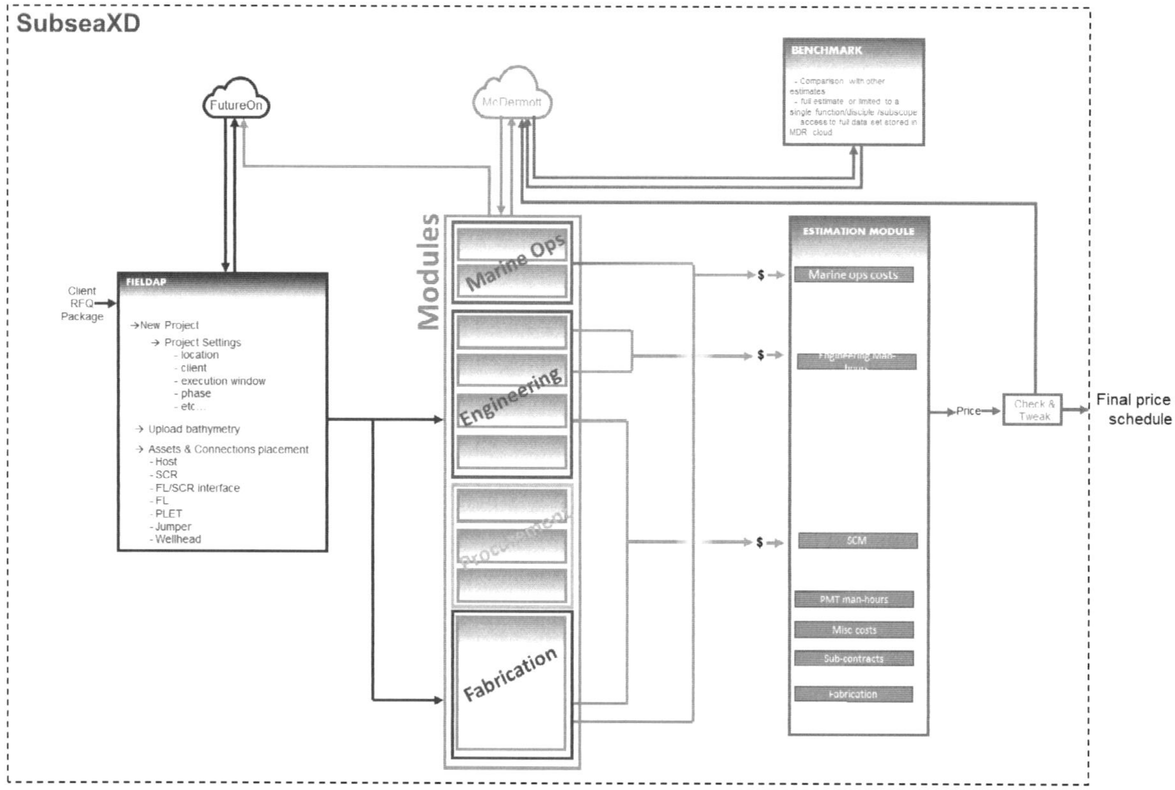

Figure 7: Estimation workflow

The blue section on the left, under "FutureOn", covers the field architecture creation process. Using client data, the team builds (or receives) a field architecture that contains the relevant assets, connections, bathymetry layers, and all main attributes that characterize these assets.

From there, various modules are used by different functions and disciplines, to make use of that data and enhance it. Importantly, all outputs coming out of the modules can be adjusted as required by the relevant discipline to fulfill the exact needs of a project.

<u>Engineering modules:</u>

For instance, the engineering department runs modules that, among other things, determine the cross-section of pipelines, the preliminary general arrangement of a Pipeline End Termination structure or the profiles of risers, as shown on Figure 8. For the latter, along with typical dimensional characteristics, the module performs code checks, calculates useful loads and automatically generates files ready to be run for in-depth dynamic finite element analysis using external software. All these modules use attributes and field data and leverage calculation scripts or simulation software running on the cloud. Not represented in Figure 7 above are the various inter-dependencies between modules within each discipline. Engineering modules not only design and further characterize the various physical assets from the field architecture, but also help generate estimates in terms of required resources (man-hours, materials/equipment, marine assets and durations) to deliver a certain scope. These man-hours are usually deliverable and asset-centric and allow engineering managers to consider multi-office execution models. This output populates the project's man-plan, consisting of required man-hours by function, title and office location, that can in turn yield an accurate cost prediction of engineering execution.

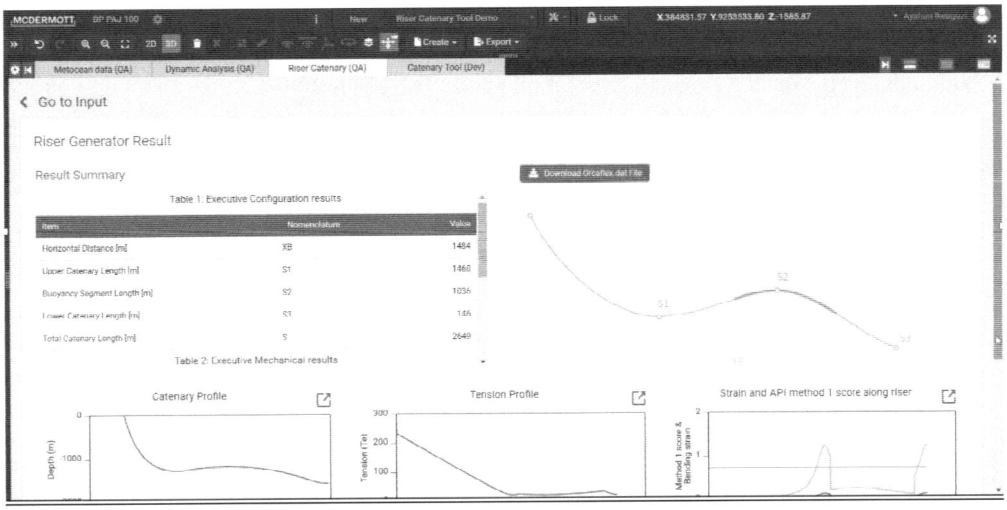

Figure 8: Example of an engineering module

Marine Operations modules:

Marine operations modules make use of both the data that resides in the digital field architecture, enhanced by the engineering modules, as well as direct outputs from engineering modules. For instance, they use cross-sectional data of pipelines and lengths to return a required number of trips, based on a chosen lay vessel, its automatically calculated cargo load, as shown on Figure 9, and installation sequence. That output then becomes the input of scheduling modules that, in turn, query a database of operational activities and uses the installation sequence and the type of pipeline terminations to generate a schedule that is representative of the prescribed installation sequence. The final output is a marine installation schedule broken down by vessel spread. Combined with the day-rate database, the marine installation costs are obtained.

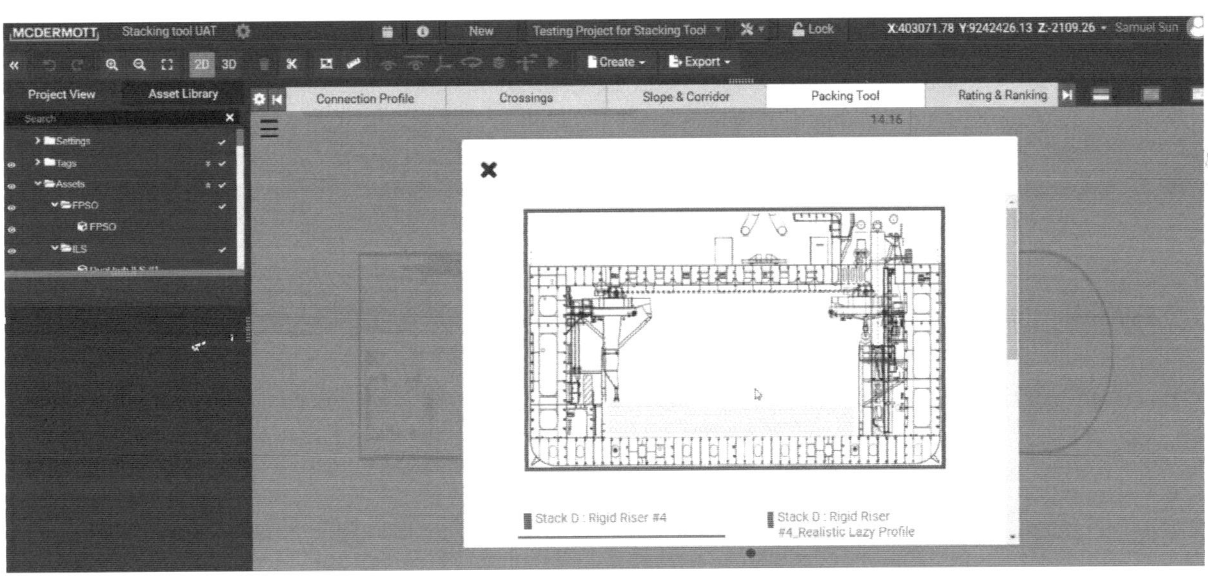

Figure 9: Example of marine operations module

Procurement modules:

The procurement list needs to be exhaustive, but it also should capture the main attributes and requirements of each item (technical and non-technical such as delivery dates). Similar to the marine operations modules, procurement modules use data residing on the FieldTwin platform that may originate from engineering modules. Algorithms iterate over the assets and their properties to return an organized equipment list, in a format that accelerates the quotation request exercise. Additionally, the equipment list with their main attributes is used to cross-reference databases of cost to quickly return a cost estimate. This greatly enhances the iterative concept definition process as comparative costs can be checked very regularly throughout development. Digitally inclined vendors can be integrated into the ecosystem to achieve a streamlined request for quotation process, without the need for human intervention.

Fabrication modules:

Mainly relevant for cost and schedule estimation purposes, these modules are crucial to ensuring reliable predictability during execution. They simulate the flowline and riser production lines using project inputs such as cross-sectional data, length, NDT (non-destructive Testing), welding and coating processes, to predict an accurate cycle time.

Forward-Looking Conclusion

Since its inception, the Subsea^{XD} ecosystem developed by McDermott, in close partnership with FutureOn, has grown in features and scope. What first started as a field architecture development tool, evolved into an automated engineering visualization and calculation tool and is today a full-blown ecosystem that encompasses all the functions involved in delivering subsea development projects with optimized solutions, increased predictability, and a shorter turnaround.

This novel ecosystem fosters collaboration both internally and externally. As the technology platform FutureOn's FieldTwin is commercially available, McDermott is not imposing a contractor-exclusive solution to their clients and vendors. This concept has so far proven successful given the many Oil and Gas companies such as Shell, Total, Schlumberger, GE Oil and Gas, and BP [Ref. 2] that have adopted it.

This data-centric approach allows McDermott to maximize the value extracted from the work performed with Subsea^{XD} on previous projects for next generation projects for an ever-improving performance. The more it is used, the faster and the better the project delivery will be.

Without a proper adoption strategy and training programs, any initiative such as this one is exposed to the risk of failure. It is therefore essential to invest in ensuring a complete buy-in to be successful in drastically transforming the ways of working.

References

[1] - Digital Transformation Initiative White Paper – Oil and Gas Industry – World Economic Forum – January 2017 (http://reports.weforum.org/digital-transformation/wp-content/blogs.dir/94/mp/files/pages/files/white-paper-2017-dti-oil-gas.pdf)

[2] – FutureOn customers (https://www.futureon.com/customers/)

INDEPENDENT ASSESSMENT OF CURRENT FLOATER CONCEPTS FOR FLOATING WIND APPLICATION

GAURAV SINGHAL
TEAM LEAD, DRILLING & FLOATING SYSTEMS, WOOD

AENGUS CONNOLLY
SENIOR CONSULTANT, WOOD

MANUEL LARANJINHA
SENIOR MANAGER, WOOD

COLIN MCKINNON
TECHNICAL DIRECTOR, WOOD

ALAN MORTIMER
INNOVATION DIRECTOR, WOOD

Proceedings of the 26th Offshore Symposium, April 6th-7th 2021, Houston, Texas
Texas Section of the Society of Naval Architects and Marine Engineers

ABSTRACT

Most of the offshore wind developments to date, globally, have been bottom-fixed foundations located in shallow waters (<30m water depth) and in close proximity to shore. However, as technology improves and as space for near-shore sites decreases, offshore wind development is projected to trend towards deeper waters. Floating wind is thus expected to become one of the leading renewable energy sources over the next decade or so. Notably, the success of pilot projects in Europe has confirmed the viability of floating wind technology, drawing in additional developers to the market. In the United States, there is a significant potential for floating offshore wind off the coast of California, Maine, and Hawaii. While the majority of current floating wind activity is concentrated in <200m water depth, further technology improvement coupled with experience from floating oil and gas developments will lead to even deeper floating wind projects in the future.

One key aspect for floating wind technology is the floater foundation that will support the wind turbine assembly. The entire unit will be moored to the seabed and be subject to challenging environment conditions throughout its service life (akin to a floating oil and gas production facility). There are several floating wind concepts currently in the market – a handful are field-proven at pilot project scale but the majority are still in development phase, each with their own unique offering. The purpose of this paper is to perform an independent qualitative assessment of the current floating wind concepts. The assessment will focus on aspects related to technology readiness, design complexity and scalability, material selection, constructability, installation, operations, and maintenance. This paper provides the offshore wind industry with an unbiased opinion on available designs as well as an insight into perceived challenges for future developments. As a disclaimer, it is noted that Wood has utilized public-domain information for this study and has no preference towards any existing floating wind concepts or designs.

Keywords: SNAME, Offshore Symposium, Floating Wind, Assessment

INTRODUCTION

The focus on renewable energy use and reduction of net greenhouse gas emissions has increased significantly in recent times. Many countries, major oil and gas operators as well as other leading industries (e.g., electrical power suppliers, manufacturing, transportation), have indicated a strong commitment towards Net Zero by 2050. To achieve this challenging objective, focus has shifted from conventional energy towards the use of renewable energy sources to supplement the global energy requirements. Offshore wind, including fixed and floating solutions, has been among the most promising renewable energy initiatives and has seen tremendous technological advancement of late. Several initiatives are underway globally to make offshore wind a commercially viable energy source compared to fossil fuels.

As technology improves further and as space for near-shore fixed wind sites decreases, floating wind is expected to become one of the leading renewable sources of energy over the next decade or so. A recent Carbon Trust report [Ref. 1] has outlined floating wind deployment across various regions of the globe using a bottom-up assessment of the floating wind market through 2040. Figure 1 shows the predicted growth of floating wind across three major regions, with over 10GW of floating wind development expected by 2030 and almost 70GW by 2040. The report marks 2030 as a significant step change in floating wind development, beyond which the rate of installed capacity is expected to increase to over 3GW/year.

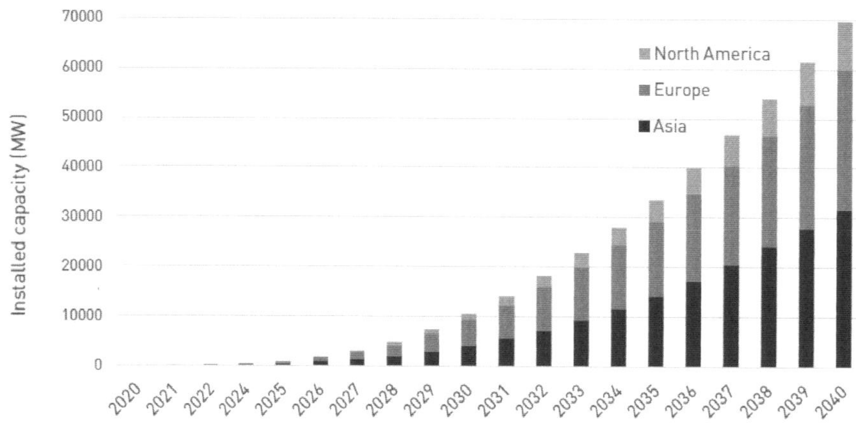

Figure 1: Floating Wind Market Trends [Ref. 1]

Aside from turbine technology, one other key aspect for floating wind development is the floater foundation that will support the wind turbine assembly. The entire unit will be moored to the seabed and be subject to challenging environment conditions throughout its service life. There are several floating wind concepts currently in the market – a handful are field-proven but the majority are still in development phase, each with their own unique offering.

The purpose of this paper is to perform an independent qualitative assessment of the current floating wind concepts. A number of key attributes are considered for the floater evaluation and each concept is rated across those attributes by use of a simple traffic light system. As an added benefit, this paper also serves as a knowledge hub with information compiled from various sources about different floater types and their salient features. The information presented provides an unbiased opinion of current industry solutions and affords prospective developers/stakeholders preliminary insight into different floating solutions for future developments.

The paper is organized as follows. "Floating Wind Substructure Types" provides a brief overview of the broad types of floating wind substructures as well as benefits and challenges of each type, and different materials of construction. "Proprietary Floating Substructure Designs" outlines the various proprietary designs across a range of substructure types. "Concept Evaluation Methodology" briefly describes the various attributes used for qualitative evaluation, followed by "Qualitative Assessment" which provides discussion and concept rankings for each substructure type. "Discussion and Concluding Remarks" summarizes and concludes the general findings from this paper.

FLOATING WIND SUBSTRUCTURE TYPES

Floating wind substructures may be broadly categorized as follows:

- Semi-submersible: Column-stabilized platform consisting of three or four columns linked by pontoons (or braces) and kept in position through a spread mooring system anchored to the seabed. The waterplanes of the spread columns provide stability whereas pontoons provide additional buoyancy. Relatively shallow draft permits quayside integration of turbine but the motions tend to be higher than some other substructure types.
- Spar: A cylindrical ballast-stabilized structure with low water plane area, with the center of gravity lower than the center of buoyancy. The simple structure of the spar is typically easy to fabricate, provides good stability and very low motions, but the large draft requirement can create logistical challenges during assembly, transportation, and installation, and can constrain deployment to waters >100m depth.
- Tension-leg platform (TLP): TLPs are anchored to the seabed with tendons (tensioned mooring lines), which provide stability as well as restrict motions. Particularly, tendon stiffness helps in minimizing the heave, roll and pitch motions of the TLP. The moderate draft and tension stability allow for a smaller and lighter structure than a spar-type. There are also challenges with the installation process and increased operational risks if a tendon fails.
- Barge: Barge has the shallowest draft among all floating substructure types. This is an advantage for installing the turbine at quayside which typically has limited water depth. However, barges tend to have greater motions in waves than semi-submersibles, which can demand more robust mooring systems. Some barge designs also include a central moonpool or catamaran configuration to provide water access as well as to spread the waterplane for improved stability.
- Hybrid designs incorporate multi-turbine floating wind platforms as well as hybrid wind-wave floating structures that harness both wind and wave energy in one system. Such designs offer higher efficiency in terms of structural tonnage per Megawatt (MW) output, but have higher design, fabrication, and installation complexity.

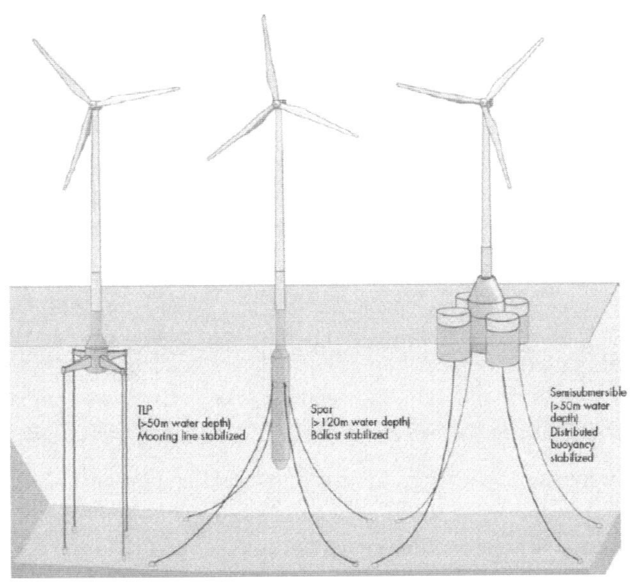

Figure 2: Floating Wind Substructure Types [Ref. 2]

Each substructure has its own advantages and disadvantages and there is no clear leader, although semi-submersible and spar designs are currently more advanced than TLP and hybrid designs. While the choice of substructure is often determined by site-specific conditions, there are some general strengths and weaknesses which are broadly acknowledged (refer to Table 1).

Table 1: Benefits and Challenges of Substructure Types

Type	Benefits	Challenges
Semi-submersible	• Not sensitive to water depth • Constructed on shore or in dry dock • Supports quayside integration and wet transport to site using conventional tugs • Significant fabrication experience globally	• Relatively higher motions compared to Spar and TLP types • Hull sizing governed by quayside integration water depths and stability requirements during integration/tow out and transport conditions
Spar	• Better motion performance than semi-submersible • Simple design and fabrication process • Unconditionally stable with center of buoyancy higher than center of gravity	• Requires deeper water depth compared to other concepts • Complex installation • Quayside turbine integration is not feasible due to deeper draft • Requires heavy lift vessels for turbine integration
TLP	• Better motion performance than semi-submersible and spars • Constructed on shore or in dry dock • Supports quayside integration and wet transport to site using conventional tugs • Significant fabrication experience globally	• Stability during transport and installation challenging • Tendon/taut mooring installation may require use of heavy lift vessels • Turbine-floater coupled dynamics may be challenging • Higher risk in case of tendon/mooring failure
Barge	• Offers shallowest draft compared to other floaters • Supports quayside integration • Simple design with significant fabrication experience	• Poorer motion performance compared to other floaters • Due to higher motions, requires larger moorings (or single-point-mooring) to restrict the floater which results in somewhat costlier mooring system
Hybrid & Multi-Turbine Designs	• Offer better efficiency in terms of structural tonnage per MW • Platform can offer more space for ancillary equipment and facilities (e.g., hydrogen production) • Hybrid wind-wave designs can increase net operability as well as total power output	• Wake effects can reduce power yield • Larger platform susceptible to higher environmental loading • More complex design, fabrication, integration, and installation • Larger overall structural mass

Most substructures have been made from steel to date. This trend stems from the fixed offshore wind and offshore oil and gas platforms, which are dominated by steel structures. However, there has been increasing interest in using concrete as an alternative material for the hull. Concrete is cheaper per unit of mass and is not exposed to high price volatility like steel, but it does increase the mass and size of the substructure compared to steel designs. In terms of fabrication, concrete fabrication yards can be established in any country, unlocking local content benefits, and potentially reducing transportation time and cost. However due to the typically larger sizes and mass compared to steel designs, concrete designs could require larger investments in port facilities to make them viable for mass production.

Steel substructures have a long history of being used offshore and steel is much easier to recycle. However, it is subject to corrosion (so is concrete via internal reinforcements, but to a lesser extent). Floating wind substructures made of steel can be assembled faster as these are not as sensitive to environmental conditions (e.g., frost and heavy rain) and are not exposed to curing time like concrete designs. Material selection will therefore depend on a number of technical and commercial considerations.

Proprietary Floating Substructure Designs

This section briefly summarizes the current proprietary floating substructure designs as well as their physical characteristics in a tabular format.

Table 2 provides a list of various floating concept designs currently being offered by a number of companies. Each of the major substructure types discussed earlier are well represented. Most designs use steel as construction material but a few concrete-based as well as combination steel-concrete designs are also available. The turbine arrangement is dominated by conventional horizontal axis turbines, reflecting current industry trends, albeit one design (SeaTwirl) offers a vertical axis turbine configuration. It is likely that other vertical axis configurations are available and also some other floater designs could also incorporate vertical axis turbines without much technical difficulty. There may be other concepts missing from this list due to lack of information available to the authors. Also, some designs have very limited information available in the public-domain and as such have been omitted from qualitative evaluation at this time.

Table 3 presents physical properties for each of the proprietary designs. These data have been compiled from Quest Floating Wind Energy's publicly available database [Ref. 3]. The authors have not independently approached any vendors to confirm the accuracy of this information. Since these data are published by Quest following their internal data acquisition procedures, the authors have assumed this information is sufficiently accurate for the purposes of qualitative assessment. Key parameters used for evaluation include minimum water depth, overall weight reported as Tons/MW, operational draft, and mooring system details.

Some basic observations are as follows:

- There is a significant variation across different designs. This is expected as the industry is still evolving and has yet to converge on a smaller number of designs which will ultimately become industry standards.
- Some designs have reached higher TRL ("Technology Readiness Level"; definitions discussed later) than others. Of the concepts considered, Hywind, WindFloat, V-shaped Semi, Mirai, and Ideol are currently the most widely deployed.
- Approximately half of the currently offered designs are based on the semi-submersible type.
- Design efficiency is improving over time. For example, both WindFloat and Ideol have lower structural weight to turbine power ratios for their more recent pilot projects than in the original demo projects.
- Concrete-based designs are heavier than comparable steel designs but involve lower cost and are easier to source.
- Most designs have a low transit draft, with the exception of Spar types, facilitating quayside integration of wind turbine (see Table 3 for installed drafts). Due to its deep draft, the spar would be towed out in a horizontal position, and then upended at site. Once ballasted down to installation draft, the wind turbine installation is conducted through use of heavy lift vessels. Specialized vessels may also be required for Hexicon and SCD Nezzy2 designs which offer multi-turbine configuration.

Table 2: Proprietary Floating Substructure Designs

S. No.	Name	Type	Owner	Material	Turbine Axis	TRL
1	Windfloat	Semi	Principle Power	Steel	Horizontal	8
2	OO Star	Semi	Olav Olsen	Concrete	Horizontal	4
3	Tetrasub	Semi	Stiesdal	Steel	Horizontal	4
4	Nautilus	Semi	Nautilus	Steel	Horizontal	4
5	EOLINK	Semi	EOLINK	Steel	Horizontal	3
6	Sea Reed	Semi	Naval Energies	Steel / Concrete	Horizontal	4
7	Moss Tripod	Semi	Moss Maritime	TBC	TBC	1
8	SCD Nezzy	Semi	Aerodyn Engineering	Concrete	Horizontal	3
9	Tri floater	Semi	Gusto	Steel	Horizontal	4
10	Pivot Buoy	Semi	X1 Wind	Steel	Horizontal	3
11	Hexicon	Semi	Hexicon	Steel	Horizontal	4
12	V shape Semi	Semi	Mitsubishi	Steel	Horizontal	7
13	VolturnUS	Semi	University of Maine	Concrete	Horizontal	5
14	Mirai	Compact Semi	Mitsui	Steel	Horizontal	6
15	MODEC	TLP/Semi	MODEC	Steel	TBC	4
16	Hywind	Spar	Equinor	Steel	Horizontal	8
17	Tetraspar	Spar	Stiesdal	Steel	Horizontal	4
18	Hexafloat	Spar	Saipem	Steel	Horizontal	4
19	SeaTwirl	Spar	SeaTwirl	Steel	Vertical	5
20	Hywind Tampen	Spar	Equinor	Concrete	Horizontal	8
21	Toda Spar	Spar	Toda	Concrete	Horizontal	6
22	Advanced Spar	Spar	IHI	Steel	Horizontal	6
23	Pelastar	TLP	Pelastar	Steel	Horizontal	4
24	SBM TLP	TLP	SBM	Steel	Horizontal	4
25	Axis TLB	TLB	Axis Energy	Steel	Horizontal	3
26	Gicon SOF	TLP	Gicon	Concrete	Horizontal	4
27	FLOW	TLP	Flowocean AB	Steel	Horizontal	1
28	Sway	Hybrid Spar/TLP	Inocean	Steel	Horizontal	3
29	BlueSath	Barge	Saitech Offshore	Concrete	Horizontal	5
30	Ideol Barge	Barge	Ideol	Steel / Concrete	Horizontal	6
31	SWACH	Circular	Sevan	Steel	Horizontal	3

Table 3: Substructure Physical Properties [Ref. 3]

Type	Name	Min Depth (m)	Tons / Megawatt	Displacement (t)	Material Wt. (t)	L (m)	W (m)	H (m)	Draft (m)	Mooring System	# Mooring Lines	Anchor Type
SEMI	WindFloat	40	478	2600	1300	50.0	40.8	24.0	14.0	Catenary	4	Drag
	WindFloat Gen2	40	273	4100	2500	55.0	55.0	30.0	20.0	Catenary	3	Drag
	WindFloat Gen3	40	298	4450	2750	75.0	75.0	30.0	20.0	Semi-taut	6	Suction
	OO Star	35	1716	15000	15000	68.0	75.0	38.0	12.0	Catenary	3	Drag, suction
	Tetrasub											
	Nautilus	35	400	-	2000	-	-	-	-	Catenary	4	Drag
	Eolink	50	-	2100	1900	66.0	58.0	50.0	20.0	Catenary	4	Suction
	Eolink Demo	50	300	1600	1500	33.0	29.0	25.0	10.0	Catenary	3	Drag
	Sea Reed	50	330	6000	3500	80.0	80.0	35.0	10.0	Catenary	6	Suction
	Sea Reed - Concrete	50	315	-	-	80.0	80.0	35.0	10.0	Catenary	6	Suction
	Moss Tripod											
	SCDNezzy - Concrete	35	133	1200	800	58.0	39.0	15.0	10.0	Turret	5	-
	SCDNezzy2 - Concrete	40	128	2040	1360	90.0	88.0	15.0	18.0	Turret	6	Drag
	Trifloater	100	360	2770	1800	76.0	87.0	30.0	13.2	Catenary	3	Drag
	Pivot Buoy	50	-	-	1500	90.0	90.0	16.0	10.0	Turret	3	Clump Weight
	Hexicon	50	580	6000	5800	180.0	80.0	20.0	15.0	Turret	10	Drag
	Hexicon 2	40	302	2450	2450	95.0	45.0	15.0	10.0	Turret	5	Suction
	V Shape Semi	50	929	6500	5000	106.0	85.0	32.0	17.0	Catenary	8	Drag
	Volturn US	40	195	-	1500	90.0	90.0	30.0	19.0	Catenary	3	Suction
	Mirai	50	1000	-	-	58.0	58.0	-	17.0	Catenary	4	Drag
	MODEC TLP											
SPAR	Hywind I	75	652	5300	1500	8.3	8.3	100.0	100.0	Catenary	3	Drag
	Hywind II	75	310	12000	2300	14.5	14.5	91.0	78.0	Catenary	3	Suction
	TetraSpar	50	334	2000	1500	45.0	50.0	38.0	70.0	Catenary	3	Drag
	Hexafloat	130	352	4350	2350	67.0	67.0	40.0	35.0	Semi-taut	6	-
	SeaTwirl 2	90	1500	-	-	100.0	7.0	100.0	70.0	Catenary	4	Suction
	Hywind III - Concrete	75	571	12000	5000	14.5	14.5	110.0	90.0	Catenary	3	Suction
	Toda Spar - Concrete	100	1125	3030	3000	7.8	7.8	91.0	76.0	Catenary	3	Drag
	Advanced Spar	75	1600	-	8000	59.0	51.0	48.0	33.0	Catenary	6	Drag
	Pelastar	75	250	1500	1174	103.0	18.0	17.0	22.0	Tethered	5	Pile
TLP	SBM TLP	50	200	-	1600	50.0	50.0	35.0	6.4	Semi-taut	6	Suction
	Axis TLB											
	Gicon	30	451	1732	1500	32.0	32.0	28.0	14.0	Tethered	4	Suction
	FLOW											
	Sway											
Barge/Misc	BlueSATH	25	755	3150	3150	35.0	25.0	12.0	7.0	Turret	6	Drag
	Ideol Barge - Steel	28	1100	12060	3100	51.0	51.0	10.0	7.5	Catenary	9	Drag
	Ideol Barge - Concrete	28	2230	6280	4360	36.0	36.0	9.5	7.5	Catenary	6	Drag
	Ideol Barge - Concrete2	28	3586	12990	10560	41.0	41.0	10.0	7.5	Catenary	9	Drag
	Ideol Barge - Concrete3	28	1603	15900	12600	50.0	50.0	12.0	8.0	Catenary	9	Drag
	SWACH											

Note: Grey shaded cells across some concepts indicate missing data

Figures 3 to 7 provide some sample images of semi-submersible, spar, TLP, barge, and multi-turbine substructure types, respectively for reference.

Figure 3: Semi-submersible examples: Windfloat (courtesy: Principle Power), Eolink (courtesy: Eolink) & OO Star (courtesy: Dr. techn. Olav Olsen)

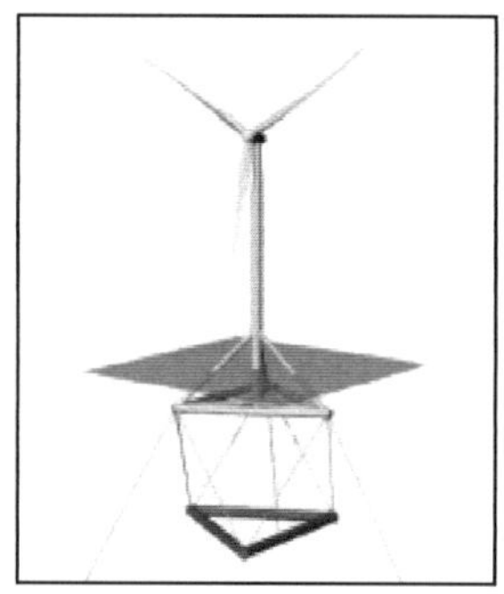

Figure 4: Spar examples: Hywind (courtesy: Equinor), Tetraspar (courtesy: Stiesdal) & Hexafloat (courtesy: Saipem)

Figure 5: TLP examples: SBM TLP (courtesy: SBM) & GICON (courtesy: GICON-SOF)

Figure 6: Barge examples: BlueSATH (courtesy: Saitec) & Ideol Barge (courtesy: Ideol)

Figure 7: Multi-Turbine example: SCD Nezzy2 (courtesy: Aerodyn)

CONCEPT EVALUATION METHODOLOGY

The evaluation methodology considers a wide range of attributes against which each concept is assessed. One of the most important attributes is the technology readiness, which is based on the TRL ("Technology Readiness Level") metric used by the floating wind industry as defined in Table 4 [Ref. 1].

Table 4: Floating Wind TRL Definition

TRL	Definition	TRL	Definition
1	Initial Concept	6	1-5 MW Demo
2	Proof of Concept	7	> 5 MW Demo
3	Numerical Modelling	8	Pilot (10-50MW)
4	Tank Testing	9	Pre-Commercial (50-200MW)
5	Scaled Testing (< 1 MW)	-	Commercial (>200MW)

Another important consideration is net carbon footprint associated with each concept (including design, fabrication, installation, operation, maintenance, and decommissioning). However, insufficient information was available to the authors at the time of writing to assess this attribute and so a neutral score is assigned to all concepts. This attribute should be considered in future studies to understand overall sustainability impact with execution of floating wind projects.

While most of the discussion in this paper is around technical considerations, a brief discussion on CAPEX is also presented. Information is primarily based on a 2015 Carbon Trust report [Ref. 4] that performed a detailed review of floating wind costs. While that report is over 5 years old and the industry has realized further cost refinements since, the data presented therein is still the most complete and current public-domain information available and considered relevant for relative comparison among different substructure types.

Figure 8 presents a relative comparison between CAPEX of different substructure types as well as a general breakdown. Each of the semi-submersible, spar and TLP types expect to deliver commercial projects at just under £2.7m/MW (~$3.72m/MW), while the hybrid options are about £2.9m/MW (~$3.99m/MW). It may be worth noting that the CAPEX differences between semi-submersible, spar, and TLP are likely within the accuracy of cost information available at this stage. Most of the concepts have therefore been allocated a neutral CAPEX score, while the hybrid and multi-turbine types have been assigned a low CAPEX score. The greatest share of the overall CAPEX comes from the turbine (41%), followed by the platform (22%), balance of system (13%, comprising electrical infrastructure, project management and contingency), installation (13%), moorings and anchors (8%), and decommissioning (3%).

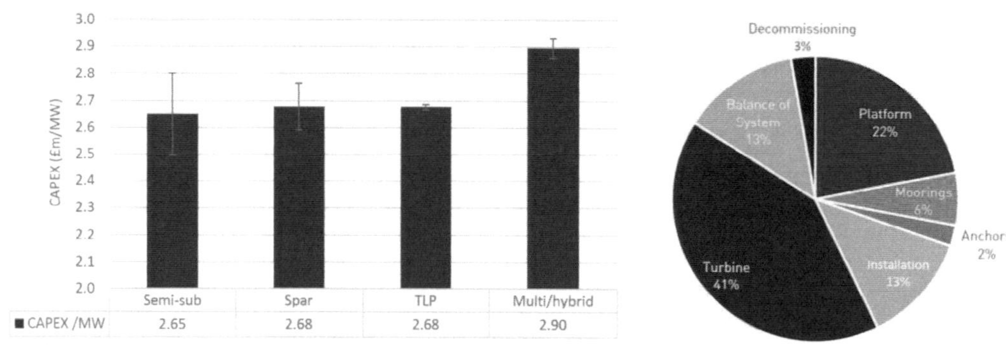

Figure 8: Commercial CAPEX for Floating Wind Substructures and CAPEX Breakdown [Ref. 4]

A summary of all attributes considered for qualitative evaluation is presented in Table 5. Each attribute has been assigned a weighting factor based on its relevance and importance. The authors appreciate that the assigned weighting factors would be open to discussion by various stakeholders on a case-by-case basis for any given floating wind development project.

Table 5: Attribute Definition and Weighting

Attribute	Weighting	Description
Technology Readiness	15%	Based on TRL; TRL 6 or more is a strength, TRL ~4-5 is neutral, TRL 3 or less is a weakness.
Design Complexity	5%	Simpler design is preferred.
Weight (Ton/MW)	8%	Weight of the hull based on Ton/MW (<300Ton/MW - strength; 300-1000 Ton/MW - neutral; >1000 Ton/MW - weakness).
Stability	5%	Considers hull stability aspects during operational, integration and installation phases.
Water Depth Limitation	8%	Whether concept is limited by water depth.
Carbon Footprint	5%	Whether the construction practices and operations of the concept will be carbon neutral. Assumed neutral across all types.

Attribute	Weighting	Description
Mooring Footprint	5%	Catenary spread will have larger footprint compared to taut and semi-taut mooring patterns; congested arrangement could cause interference issues for larger developments.
Extreme Weather Tolerance	5%	Overall robustness of the concept in extreme weather conditions during operational as well as temporary conditions (installation, hook-up, tow).
Onshore/Quayside Integration	8%	Considered a strength if the concept can be integrated at quayside.
Constructability	5%	Constructability depends on design simplicity and yard experience.
Installability	8%	Considered a strength if the concept can be installed with typical installation vessels.
Scalability	8%	Addresses scalability of the concept to multiple units.
CAPEX	10%	Limited data and therefore considered for major substructure types based on published information [Ref. 4].
Maintenance/Decommissioning	5%	Considered a strength if maintenance aspects are simple. Decommissioning considers whether the installation process can be easily reversed.
Total	**100%**	

In order to rank each concept across the above attributes, a simple traffic light system has been implemented as shown in Table 6. Green (score of 5) is considered a strength, yellow (score of 3) is considered neutral, and orange (score of 1) is considered a weakness.

Table 6: Qualitative Ranking Criteria

Classification	Score
Strength	5
Neutral	3
Weakness	1

Qualitative Assessment – Semi-Submersibles

Qualitative assessment results for concepts based on semi-submersible substructure type are presented in Table 7. In total 12 concepts have been evaluated and compared. Multi-turbine designs (e.g., SCD Nezzy2 and Hexicon) are based on semi-submersible type and are also included in this discussion. For three designs (Tetrasub, Moss Tripod and MODEC), no public-domain information is available and therefore those were excluded from this assessment.

Most semi-submersibles have a simple design with no depth limitation (except during quayside integration, which is a concern for other types also) and offer reliable extreme weather tolerance as motion response is larger compared to Spar and TLP. Semi-submersibles are also easy to construct with fewer construction sequences and several yards have experience building semi-submersibles. Installation is relatively straightforward and the concept can be easily scaled to support large turbine sizes. As a result of these characteristics, most semi-submersible concepts are allocated a higher ranking in the current evaluation.

Table 7: Qualitative Ranking: Semi-submersibles

Attribute	Weighting	Windfloat	OO Star	Nautilus	EOLINK	Sea Reed	SCD Nezzy	Tri floater	Pivot Buoy	Hexicon	V shape Semi	VolturnUS	Mirai
Technology Readiness	15%	5	3	3	1	3	1	3	1	3	5	3	5
Design Complexity	5%	5	3	3	3	5	3	5	3	5	5	3	5
Weight (Ton/MW)	8%	3	1	3	5	3	5	3	5	3	3	5	3
Stability	5%	5	5	5	5	5	3	5	5	5	5	5	5
Water Depth Limitation	8%	5	5	5	5	5	5	5	5	5	5	5	5
Carbon Footprint	5%	3	3	3	3	3	3	3	3	3	3	3	3
Mooring Footprint	5%	5	5	5	5	3	3	5	5	1	1	5	5
Extreme Weather Tolerance	5%	5	5	5	3	3	3	3	3	3	3	5	3
Onshore/Quayside Integration	8%	5	5	5	5	5	3	5	5	3	5	5	5
Constructability	5%	5	5	3	5	5	3	5	5	3	5	5	5
Installability	8%	5	5	5	5	5	3	5	5	3	5	5	5
Scalability	8%	5	5	5	5	3	3	5	5	3	5	5	5
CAPEX	10%	3	3	3	3	3	1	3	3	1	3	3	3
Maintenance/Decommissioning	5%	5	5	5	5	3	1	3	5	1	5	5	5
Total Score	**100%**	**4.54**	**3.98**	**4.04**	**3.90**	**3.78**	**2.72**	**4.04**	**3.90**	**2.96**	**4.24**	**4.30**	**4.44**

Some specific observations are noted below:

- WindFloat, V-shape Semi and Mirai have attained the highest TRL levels to date, all with successful offshore pilot projects.
- Structural weight (Ton/MW) varies among the different designs due to material choice (steel vs. concrete) as well as the semi-submersible configuration. OO Star (based on concrete) is the heaviest among all concepts and therefore evaluated, in relative terms, as a weakness. However, it is easier to construct concrete-based designs anywhere in the world, enabling local content and potentially saving overall CAPEX during execution.
- Semi-submersibles are overall quite stable units during operation. Although, temporary condition such as those experienced during integration and installation could pose challenges, particularly with multi-turbine concepts (i.e., SCD Nezzy and Hexicon).
- All semi-submersible designs will typically use a catenary mooring system offering increased ability to cope with extreme weather events (compared to taut or semi-taut moorings). However, the mooring footprint is typically 8-10 times the water depth (the factor would reduce in deeper waters) and could therefore impact the overall field layout for larger numbers of turbines. Most designs considered in Table 7 have 3-6 mooring lines but some designs have 8-10 mooring lines and are therefore assigned a weakness for this attribute.
- Maintenance, including shore-based that involves disconnection and tow to shore, is expected to be simple for most concepts, but it is likely that multi-turbine platforms (i.e., Hexicon and SCD Nezzy2) will involve a more complex return to port for maintenance.
- In summary, the top five semi-submersible concepts based on the assessment above are as follows:
    - Windfloat (total score 4.54)
    - Mirai (total score 4.44)
    - VolturnUS (total score 4.30)
    - V-shape Semi (total score 4.24)
    - TriFloater & Nautilus (total score 4.04)

  Other notable concepts are OO Star (concrete-based), Pivot Buoy, and Eolink.

QUALITATIVE ASSESSMENT – SPARS

Qualitative assessment results for concepts based on spar substructure type are presented in Table 8. In total 8 spar concepts have been evaluated and compared. Some key observations are noted below:

- Spar concepts have the simplest design (single cylinder) and offer very good motion performance (particularly small heave and pitch).
- Hywind, Hywind Tampen (concrete based), Toda Spar and Advanced Spar have the highest TRL levels among all selected concepts.
- Most conventional spars are heavy structures with deep draft. However, some innovative concepts offer weight advantages (e.g., Hexafloat, SeaTwirl) as well as shallower draft (e.g., Advanced Spar, Pelastar).

- Spar can be constructed easily, however the turbine integration and installation can be challenging and pose higher risk. Because of this reason, the overall ranking of spars is lower than semi-submersibles.
- In summary, the top three spar concepts are as follows, but most spar concepts have similar ranking:
  - Tetraspar (total score 3.72)
  - Advanced Spar (total score 3.66)
  - Hywind (total score 3.54)

Table 8: Qualitative Ranking: Spars

Attribute	Weighting	Hywind	Tetraspar	Hexafloat	SeaTwirl	Hywind Tampen	Toda Spar	Advanced Spar	Pelastar
Technology Readiness	15%	5	3	3	3	5	5	5	3
Design Complexity	5%	5	3	3	5	5	5	3	3
Weight (Ton/MW)	8%	3	3	3	5	3	1	1	3
Stability	5%	5	5	5	5	5	5	5	5
Water Depth Limitation	8%	1	1	3	1	1	1	5	5
Carbon Footprint	5%	3	3	3	3	3	3	3	3
Mooring Footprint	5%	5	5	3	5	5	5	3	3
Extreme Weather Tolerance	5%	5	5	5	5	5	5	5	3
Onshore/Quayside Integration	8%	1	5	3	1	1	1	3	1
Constructability	5%	5	5	5	5	5	5	3	3
Installability	8%	3	5	3	3	3	3	3	3
Scalability	8%	5	5	5	5	5	5	5	5
CAPEX	10%	3	3	3	3	3	3	3	3
Maintenance/Decommissioning	5%	1	3	3	3	1	3	3	3
Total Score	**100%**	**3.54**	**3.72**	**3.46**	**3.50**	**3.54**	**3.48**	**3.66**	**3.26**

QUALITATIVE ASSESSMENT – TLPS & BARGES

Since there are fewer TLP and barge/miscellaneous designs, qualitative assessment results for these substructure types are combined and presented in Table 9. Three TLP concepts and three barge/miscellaneous concepts have been evaluated and compared. Two TLP concepts (i.e., FLOW and Sway TLP) have very limited information available in the public-domain and therefore have been excluded from current evaluation.

Some key observations are noted below:

- The current TLP designs for floating wind (particularly, SBM TLP and Axis TLB) use a much lighter hull structure to support the wind turbine.
- Barge type solutions have simpler design but are prone to higher wave loads.
- Ideol barge has the highest TRL, while Axis TLB (Tension Leg Buoy) and SWACH have the lowest TRL
- TLPs are moored via tendons or taut-moorings that reduce the overall mooring footprint but is typically more costly than a spread mooring. However, the installation may require use of specialized vessels.
- SBM TLP claims to be the lightest structure on the market and has a strong heritage in hull designs from oil and gas experience.
- Axis TLB is another lightweight structure and comprises of proprietary gravity anchor, tensioned tendons, and buoyancy tanks to provide the necessary stability. The concept has only been tank tested and therefore has a low TRL. The gravity-based anchors could pose challenges in hard soils.
- Gicon SOF uses a 4-column hull structure made of pre-stressed ultra-high-performance concrete pipes. As such, the hull structural weight is larger than other TLP concepts.
- In summary, the top three TLP or Barge concepts are as follows:
  - Ideol Barge (total score 4.18)
  - SBM TLP (total score 4.10)
  - Axis TLB & Gicon SOF (total score 3.94).

Table 9: Qualitative Ranking: TLPs & Barges

		TLP Type			Barge / Misc		
Attribute	**Weighting**	SBM TLP	Axis TLB	Gicon SOF	BlueSATH	Ideol Barge	SWACH
Technology Readiness	15%	3	1	3	3	5	1
Design Complexity	5%	5	5	3	3	5	3
Weight (Ton/MW)	8%	5	5	3	3	1	3
Stability	5%	5	5	5	5	5	5
Water Depth Limitation	8%	5	5	5	5	5	5
Carbon Footprint	5%	3	5	3	3	3	3
Mooring Footprint	5%	3	5	5	3	1	3
Extreme Weather Tolerance	5%	3	5	5	3	5	5
Onshore/Quayside Integration	8%	5	5	5	5	5	5
Constructability	5%	5	5	3	3	5	3
Installability	8%	5	5	5	5	5	5
Scalability	8%	5	3	5	5	5	5
CAPEX	10%	3	3	3	3	3	3
Maintenance/Decommissioning	5%	3	3	3	3	5	3
Total Score	**100%**	**4.10**	**3.94**	**3.94**	**3.74**	**4.18**	**3.54**

DISCUSSION & CONCLUDING REMARKS

This paper presents findings from a desktop study that evaluated a number of different floating wind concepts. The qualitative assessment used a range of attributes, each with assigned weighting factor, to compare different solutions through a simple traffic light system. While the TRL attribute was assigned the highest weighting factor in our evaluation, concepts which score low on the TRL scale are not necessarily weak designs. On the contrary, innovative designs have the potential to reduce costs and assist in driving down the overall levelized cost of energy. Relatively speaking, floating offshore wind is still in its early stages, and there are several promising novel concepts (e.g., Eolink, Sea Reed, Pivot Buoy, Hexafloat, Axis TLB, SATH) that can help with reducing the energy costs significantly.

In conclusion, the following aspects are noted:

- **Leading concepts**. Concepts based on semi-submersible type (i.e., Windfloat, Mirai, VolturnUS, V-shaped Semi) appear to be the leading designs given their high overall ranking. All have reached advanced TRL and have gained industry confidence in their respective designs. OO Star offers a concrete alternative, should a traditional semi-submersible be preferred over other designs and if local content is important (concrete fabrication yards can be setup anywhere). Ideol barge also has a high ranking due to its advanced TRL; while SBM TLP, with the lightest structure on the market, is among the leading TLP concepts.
- **Strong contenders**. Innovative designs such as Eolink, Sea Reed, Pivot Buoy, Hexafloat, Axis TLB, and SATH also look promising. They have obvious benefits over more traditional designs, and they will undoubtedly develop further over the coming years. If the demo projects can be successfully followed by pre-commercial arrays, any of these would merit serious consideration.
- **Concepts that need further work**. Multi-turbine concepts, such as Hexicon and SCD Nezzy, look quite promising as they have the highest energy output per ton of weight. While both concepts have shown improvements, they remain unproven and carry significant overall risk during design, integration, and installation. These concepts need further work for industry to gain more confidence in multi-turbine designs.

ACKNOWLEDGEMENTS & DISCLAIMER

The authors are grateful to the many organizations who have openly shared information, most notably Quest Floating Wind Energy and the Carbon Trust. The R&D support from Wood plc. to undertake this study is also gratefully acknowledged. As a disclaimer, it is noted that Wood has utilized public-domain information for this study and has no preference or affiliation with any existing floating wind designs at the time of writing.

REFERENCES

1. Spearman, D., Strivens, S., Matha, D., Cosack, N., Macleay, A., Regelink, J., Patel, D. and Walsh, T., 2020. Floating Wind Joint Industry Project: Phase II Summary Report. The Carbon Trust.
2. Global Wind Energy Council (GWEC), 2020. Global Offshore Wind Report 2020. [Online at https://gwec.net/global-offshore-wind-report-2020/].
3. Quest Floating Wind Energy, 2020. Floater Concepts. [Online at https://questfwe.com/concepts/].
4. James, R. and Ros, M.C., 2015. Floating offshore wind: market and technology review. The Carbon Trust.

A FLEXIBLE RISER DIGITAL TWIN FOR REAL-TIME STRUCTURAL INTEGRITY MANAGEMENT

GABRIEL ROMBADO
TECHNICAL ADVISOR, EXXONMOBIL GLOBAL PROJECTS, HOUSTON, TX (USA)

HEMA WADHWA
DIGITAL & ENGINEERING SPECIALIST, INTECSEA/WORLEY, PERTH, AUSTRALIA

NATHAN COOKE
ENGINEERING SPECIALIST, INTECSEA/WORLEY, ST. JOHN'S, CANADA

DHARMA PASALA
ENGINEERING SPECIALIST, INTECSEA/WORLEY, HOUSTON, TX (USA)

MIKE PAULIN
OPERATIONS DIRECTOR, INTECSEA/WORLEY, ST. JOHN'S, CANADA

PUSHKAR RAO
HEAD OF DATA REFINERY, WORLEY, CALGARY, CANADA

ARYA MAJED
TECHNICAL ADVISOR, INTECSEA/WORLEY, HOUSTON, TX (USA)

Proceedings of the 26th Offshore Symposium, April 2021, Houston, Texas
Texas Section of the Society of Naval Architects and Marine Engineers

ABSTRACT

A real-time understanding of the structural integrity of flexible risers is critical to avoiding failures as well as unnecessary and costly replacements. Accurate and efficient computation of tensile armor wire stress time-histories is the key to enabling the implementation of a flexible riser digital twin. Accuracy requires capturing the nonlinear kinematics of the flexible's helically contra-wound tensile armor layers and their interaction with the other metallic and thermo-plastic layers in a dynamic simulation. While it is generally accepted that high-fidelity 3D finite element models best capture the complex kinematics required to produce accurate stresses, the main issue is that simulation time remains prohibitive, for example, a single bending cycle on modern day multi-core computers typically requires 24-48 hours. Therefore, any successful implementation of a flexible riser digital twin requires an accelerated solver capable of real-time computation of 1000s of bending cycle inputs into the high-fidelity finite element model. This paper demonstrates a flexible riser digital twin applications centered around the accelerated solver FLEXAS.

Keywords: Flexible Risers, Digital Twin, Integrity Management, Accelerated Solvers, Finite Element Model

INTRODUCTION

Flexible risers are inherently complex and highly nonlinear systems. Stick/slip friction of the tensile armor wires (Figure 1) coupled with geometric nonlinear behavior of the riser dominate the wire kinematics. A high-resolution understanding of these nonlinearities is required for accurate wire stress predictions. Tensile armor wires are the most fatigue critical components of the flexible riser. A tensile armor wire at a critical cross section that physically slips later than predicted by an inaccurate model/simulation can have a significant impact on fatigue life predictions. Furthermore, tensile armor wires may be corroded/damaged by flooded annulus conditions or corrosive gas diffusion which compounds the challenge of accurately capturing the wire kinematics. These are the critical fatigue life cases where the model/simulation must possess the fidelity for accurate predictions.

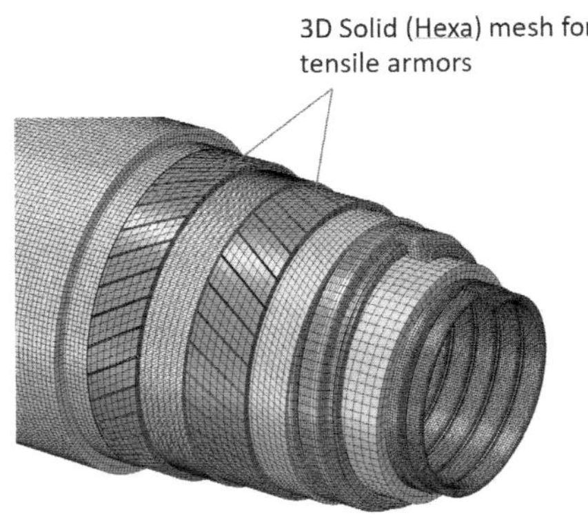

Figure 1 – A 3D High-fidelity Flexible Riser Finite Element Model Utilized to Depict Helically Wound Tensile Armor Wires

Fatigue of flexible risers due to tensile armor wire failures have long been an industry wide problem. Four known instances of tensile armor fatigue failure occurred between 2017 and 2019 (Figure 2). Initial indications suggest that there are large numbers of flexible risers under similar conditions. Replacement cost of a single flexible riser is on the average 20M USD and the cost of production loss can be several multiples greater.

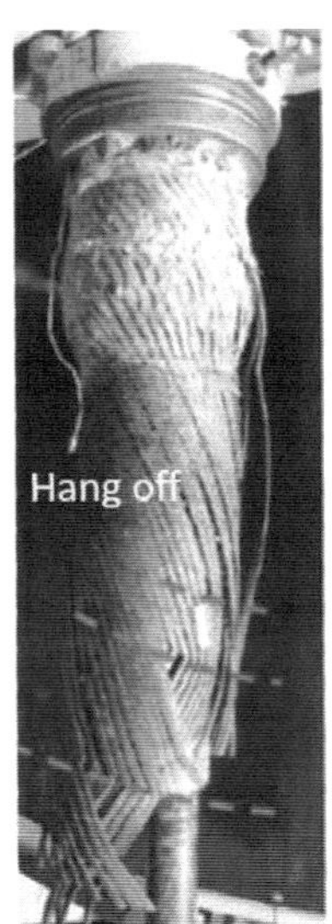

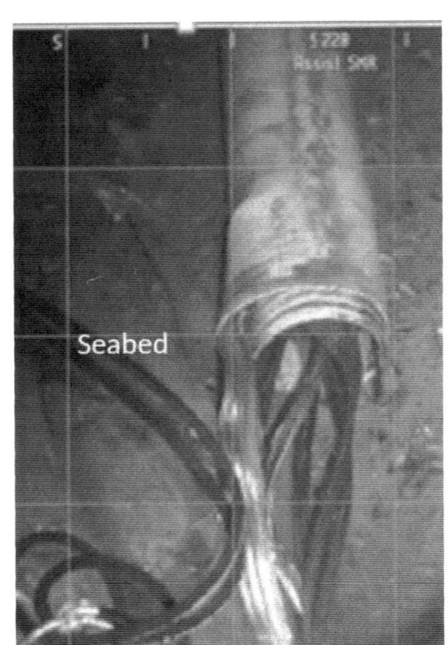

Figure 2 – Fatigue Failure of a Flexible Riser at the Hang-off due to Broken Tensile Armor Wires (2019); Remaining Riser on the Seabed

A real-time understanding of the structural integrity of flexible risers is critical to avoiding failure, unnecessary replacements, production downtimes, and significant associated costs. However, accurate predictions of flexible riser integrity and fatigue damage remains a major challenge, especially considering the potential for degraded conditions such as flooded annulus and tensile armor wire damage/corrosion. Efficient computation of time-consistent stresses for accurate utilization factors and remnant fatigue life calculations requires capturing the kinematics of the flexible's helically contra-wound tensile armor layers and their interaction with the other metallic and thermo-plastic layers in a dynamic simulation. It is widely accepted that high-fidelity 3D Finite Element (FE) models are the best predictors of the complex nonlinear (stick-slip friction) kinematics and accurate stresses.

One main issue preventing the development of a flexible riser digital twin has been the computational speed of commercially available solvers. *Perdrizet et al (2012) provide simulation times for a Technip/IFP flexible riser (7 meter span) for 1 bending cycle to be 48 hours on 36 CPUs running in parallel.* Similar computation times have been realized by the authors at ExxonMobil and Worley. This poses a major computational constraint to the development of a flexible riser digital twin which requires the near real-time computations of 1000s of bending cycles. As such, alternate solutions involving interpolation/post-processing and machine learning have been proposed. However, these solutions involve major drawbacks addressed in a later section of this paper.

The advent of accelerated solvers in the O&G industry has provided an enabling technology for the efficient computation of detailed large order FE models. This includes real-time computation of flexible riser irregular wave tensile armor wire stress time-histories. These more accurate predictions of stress and fatigue states make advanced integrity management of flexible risers inclusive of damage assessments, life extension, and digital twins practical where accelerated solvers play a critical role.

This paper is focused on the construction of a flexible riser digital twin via a full demonstration problem. The main building blocks for a flexible riser digital twin (focusing on hang-off section) include (i) detailed high-fidelity FE model of the flexible riser/bend stiffener for the hang-off section (ii) a 6 degree of freedom (DoF) set of inertial measurement units located on the FPSO at the riser hang-off location, (iii) force transducer measurements for hang-off tension, and (iv) an accelerated numerical solver. The objective is to enforce the sensor data and tension onto the FE model and compute via accelerated solver real-time tensile armor stresses which will be post-processed into utilization factors and remnant fatigue life. This framework can also make use of inspection data including annulus condition monitoring and tensile armor wire damage. With this, the operator can assess real-time utilization and fatigue life data to make timely and informed decisions relative to inspections and life extension.

FLEXIBLE RISER FINITE ELEMENT MODEL

Figure 3 depicts the evolution of the flexible riser FEM from 1990s (left) to present day (right). Simplified FEMs idealize the complex geometry and must introduce numerical "dials" to recover lost behavior. This results in lost accuracy and limited predictive capabilities. Furthermore, since the model is nonlinear, the computation times of commercial solvers remain long. As such, this simplified model is not congruent with a real-time integrity management /digital twin capability. Detailed FEMs (right) represent the precise 3D geometry and load paths while requiring no additional assumptions. With this, it captures the complex wire kinematics and interactions with other metallic and thermoplastic layers resulting in accurate wire stresses. Additionally, it can also capture damaged wire kinematics – an important requirement for a digital twin.

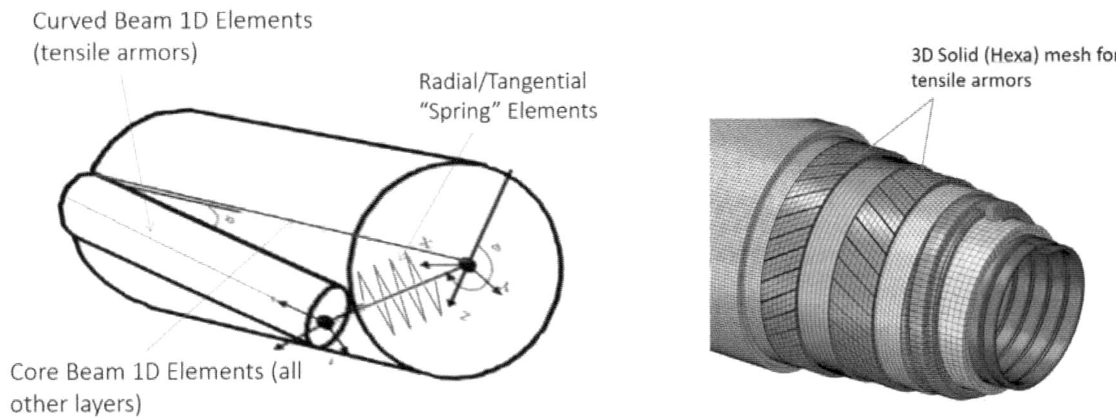

Figure 3 – Evolution of the Flexible Riser FEM from Low Fidelity (left) to High-fidelity (right)

This section constructs the digital twin model for the hang-off. A single pitch FEM (Figure 4) of a flexible riser for an ExxonMobil study is utilized for demonstration as well as providing the basis for the multi-pitch length hang-off model construction.

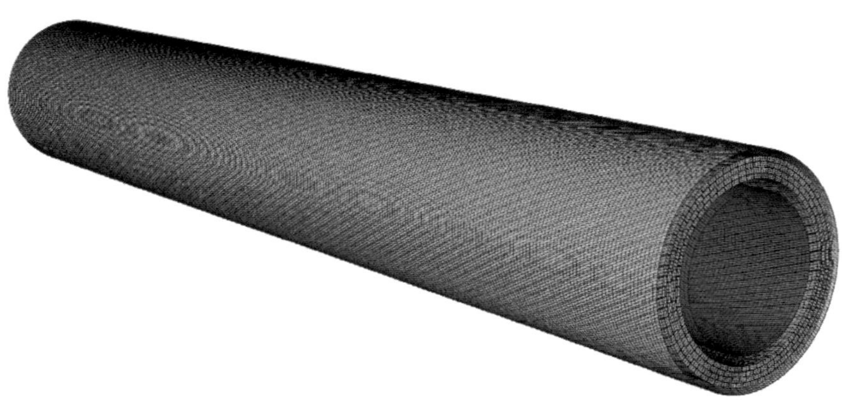

Figure 4 –Single Pitch Length FEM of a Flexible Riser

This four tensile armor layer flexible has a 9.5" ID with 1 pitch length equating to 2.89 m. The FEM involves 2.35M nodes, 1.33M elements, and a total of 14.1M degrees of freedom (DoF). Figure 5 provides a cutaway to clearly show the four tensile armor wire layers, modeled with hexa solid elements and anti-wear layers.

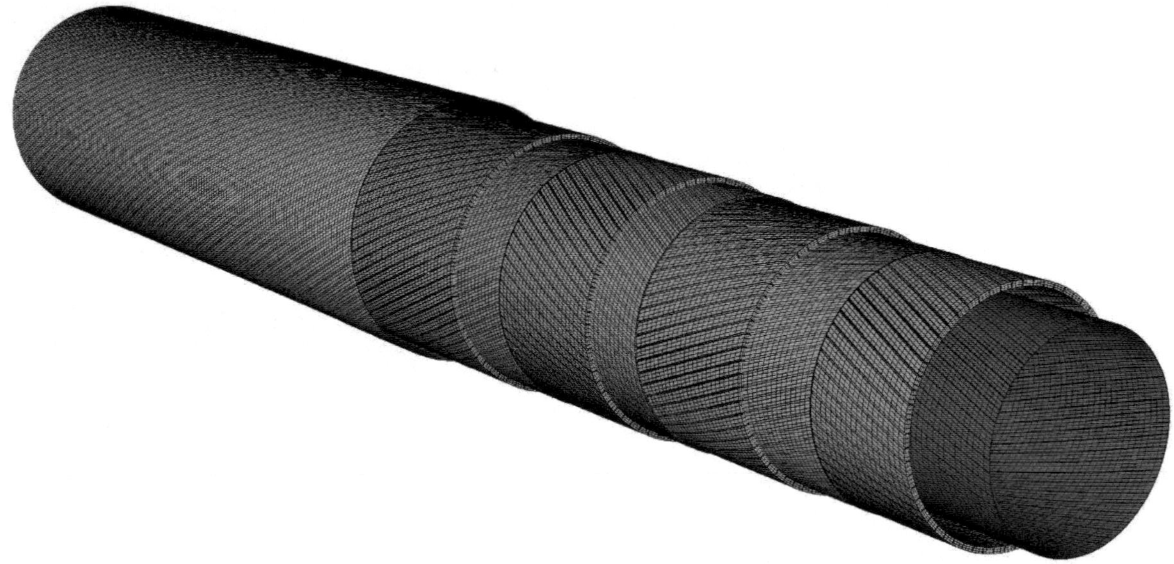

Figure 5 – Flexible FEM Cutaway Showing the Tensile Armor and Anti-wear layers

Replicating the 1-pitch length FEM 9 times provides the 10 pitch length (28.88 m) hang-off section which represents 141M unreduced DoF. Figure 6 details the final hang-off flexible FEM.

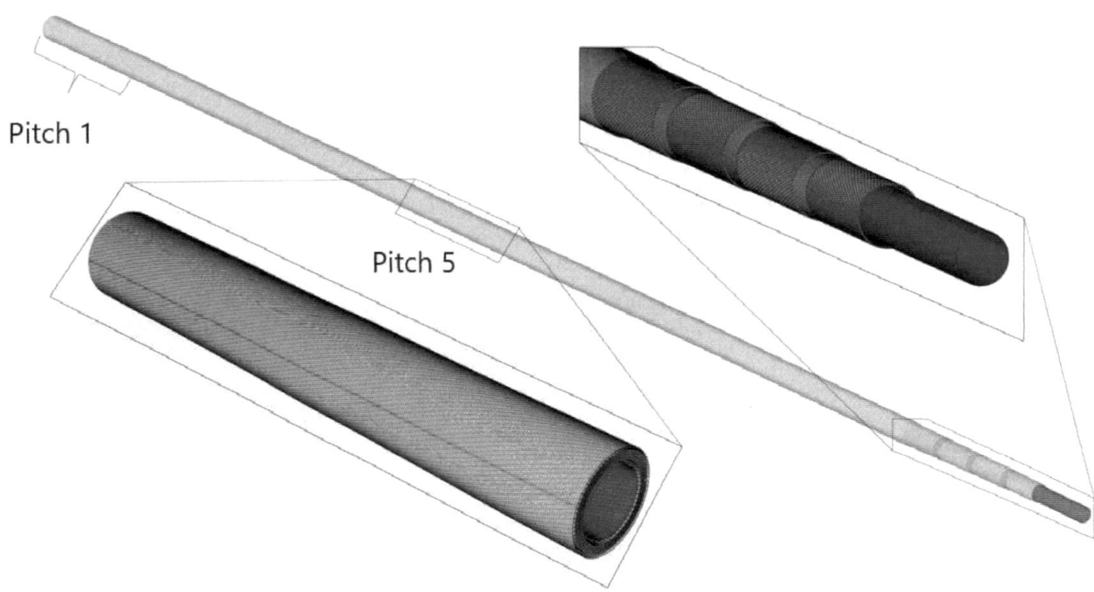

Figure 6 – 10 Pitch (28.88 m) Flexible Riser FEM

ANALYSIS CONFIGURATION AND ACCELERATED SOLVER DEMONSTRATION

Figure 7 provides the analysis configuration under time-varying cyclic loading. The above digital twin is activated by long duration irregular wave 2-axis bending rotations and tension time-histories captured by the 6 DoF inertial measurement unit and force transduce, respectively. A basic checkout of the analysis model is executed here for an enforced rotation of 4 degrees varying sinusoidally at a 5 second period for 10 seconds. A wall tension of 4060 kN and internal pressure of 5.2 MPa are utilizing to demonstrate the wire stiction for stick/slip state transition. Figure 8 provides the bending moment at the hang-off due to the sinusoidal (regular wave) inputs.

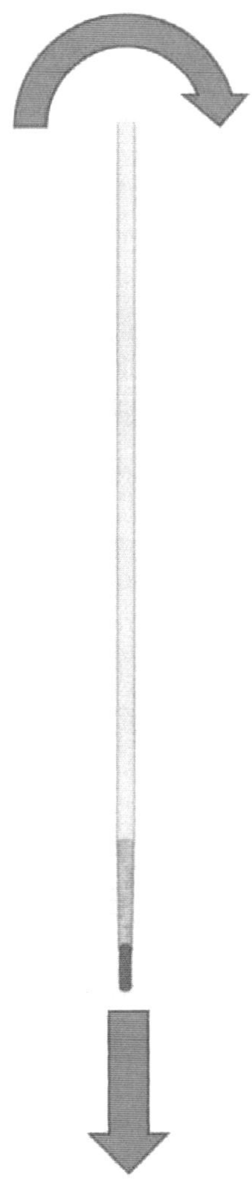

Figure 7 – Hang-off Analysis Configuration

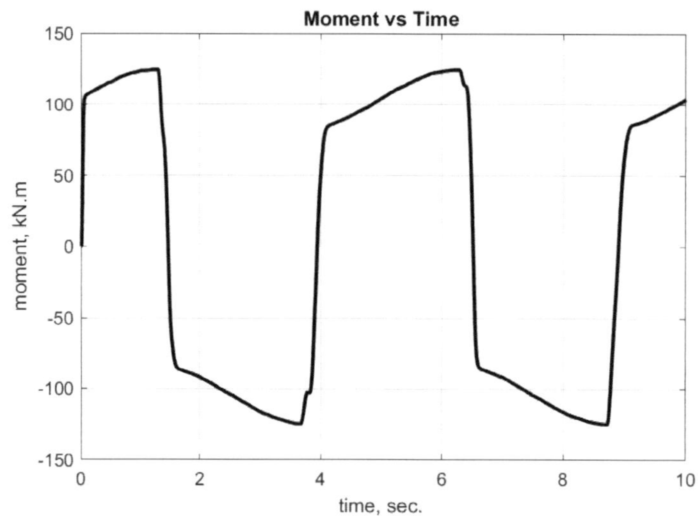

Figure 8 – Hang-off Bending Moment Time-History Demonstrating Stick/Slip Behavior Due to Tensile Armor Wire Friction

Some observations are in order. (1) Unlike a steel riser, the flexible riser does not produce a sinusoidally varying (linear) bending moment time-history response to an input sinusoidally varying rotation. Instead, the simulation resulted in a highly nonlinear output moment behavior due to the stick/slip characteristics of the tensile armor wires. (2) The nonlinear time-history, the resultant of the summation of the wire stresses times their respective moment arms to the neutral axis, displays a very rapid buildup of moment to approximately 100 kN.m prior to the flexible initiating slip in order to relieve the buildup of strain energy. It is precisely this design feature of the flexible which enables it to survive the harsh environment fatigue requirements unlike steel risers. (3) these nonlinear dynamic simulations were executed practically in real-time in the accelerated solver FLEXAS (for accelerated solver theory and O&G industry verification & validation, see Majed et al 2016 and 2017).

To better understand the basic features of the nonlinear behavior, the moment time-history can be plotted against the kinematic variable, in this case, curvature. Figure 9 provides this depiction.

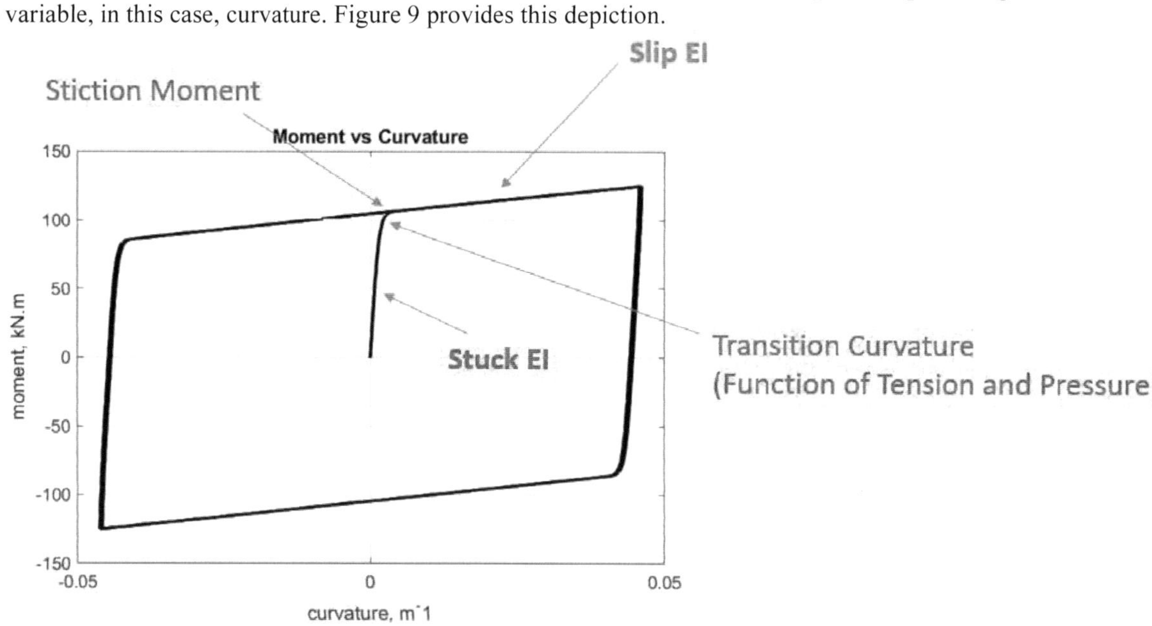

Figure 9 – Hang-off Bending Moment Time-History Plotted Against Curvature

Figure 9 clearly reveals the stick/slip characteristics of the subject flexible riser. From this, the "stuck" and "slip" bending stiffness (EI) can be extracted. For the subject flexible with four tensile armor layers, it can be calculated that the ratio of the stuck to slip stiffness is ~150.

To further demonstrated the speed of the accelerated solver, the subject flexible riser nonlinear dynamic simulation is now executed with a top rotation period of 1 seconds for a total of 50 seconds. The resulting hang-off nonlinear moment and hysteresis are provided in Figures 10 and 11, respectively.

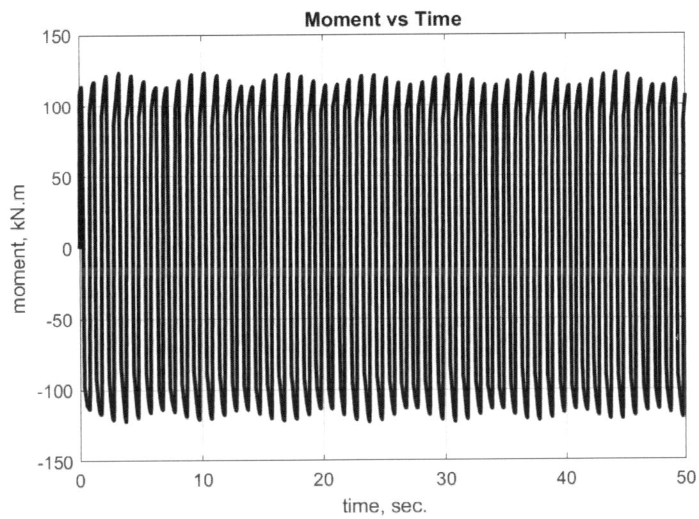

Figure 10 – Hang-off Bending Moment Time-History (50 seconds at a 1 second period)

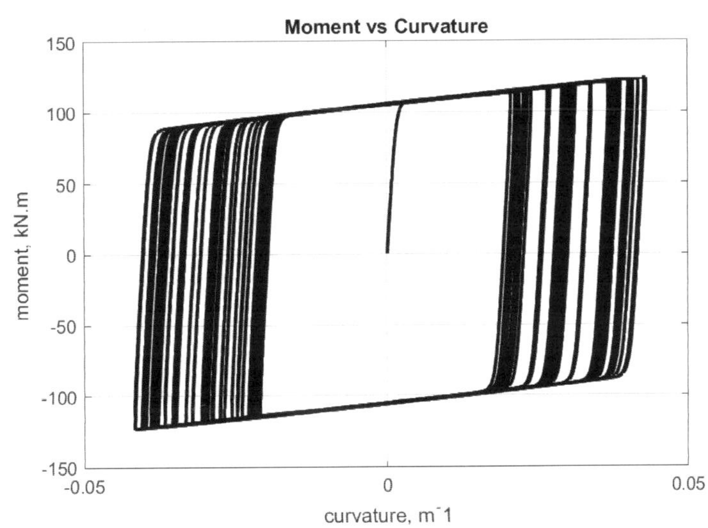

Figure 11 – Hang-off Bending Moment Time-History Plotted Against Curvature (50 seconds at a 1 second period)

Hysteretic characteristics of the flexible are expected to vary span wise. Figure 12 provides the hysteretic moment 4 pitches down from the hang-off.

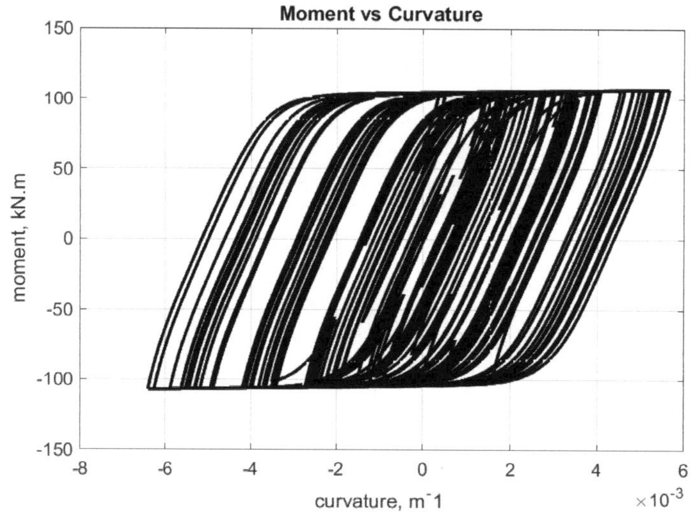

Figure 12 – Bending Moment Hysteresis Characteristics 4 Pitches Down from Hang-off (50 seconds at a 1 second period)

It is important to note that the FLEXAS accelerated solver has provided the industry's first simulation-based solution to the irregular wave flexible fatigue problem. This involved 1-hour long tension/curvature time-histories with 1000s of irregular wave bending cycles input into a detailed 3D flexible FEM. Figure 13 presents a comparison of irregular wave and regular wave tensile armor wire fatigue spectra at the critical wire location, which contained the highest cumulative fatigue damage. At this location, the fatigue damage ratio was found to be 6.8, i.e. the fatigue life calculated with irregular waves was about 7 times greater than the fatigue life computed with regular waves for a client flexible described in Majed et al (2017).

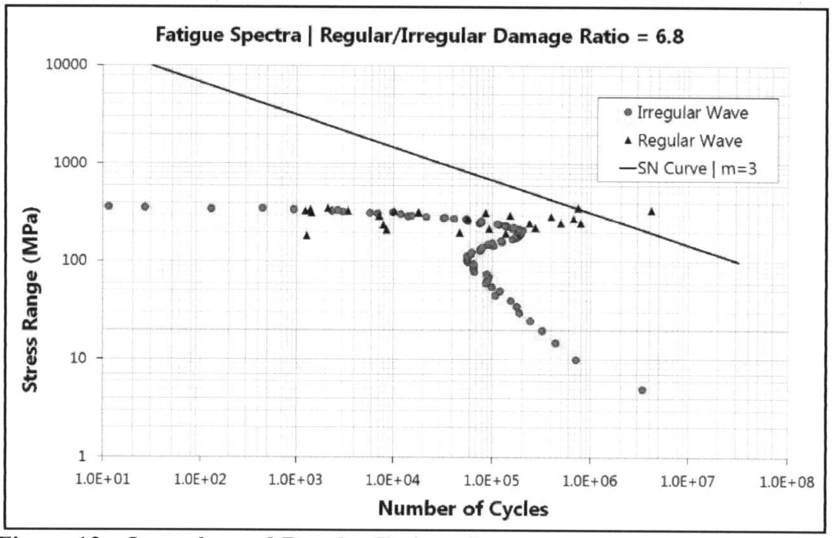

Figure 13 – Irregular and Regular Fatigue Spectra at Critical Wire Location

With reference to Figure 13, the irregular wave fatigue spectrum shows a more realistic distribution and greater resolution. As expected, the highest stress ranges occur infrequently, and are limited to a relatively low number of cycles. This coincides with the largest waves being relegated to the lower probability sea-states. More importantly, the spectrum exhibits a sharp drop-off in stress amplitudes for higher cycle counts. This limits the fatigue damage caused by high-frequency (low period) smaller wave load components. This is precisely the level of realism and accuracy that is required for real-time structural integrity management and remnant fatigue life calculations.

DAMAGED ARMOR WIRES

Tensile armor wires can incur damage or even break for a number of reasons including higher than anticipated loads and corrosion due to flooded annulus conditions and gas diffusion. Damaged tensile armor wires can be detected via flexible riser inspection systems. It is imperative for real-time integrity management system to be able to resolve the impact of damaged wires on stresses and remnant fatigue life. In this section, it is demonstrated that the high-fidelity 3D FEM does possess the accuracy requirement to capture damaged wire kinematics and stresses.

In a series of experiments conducted by deSousa et al (2015), tensile armor wires for a number of broken wire configurations were extracted. Doynov et al (2020) utilized these results to (1) high-fidelity 3D FEM accuracy in capturing damaged wire kinematics and strains, (2) performance of accelerated solver in resolving the subject kinematics and strains. Figure 14 provides one such flexible riser FEM inclusive of a specific damage configuration utilized in this study.

Figure 14 – Finite Element Model of One Configuration of Flexible Riser with Broken Wires

Figure 15 provides the integrated FEM for the full test configuration (approximately 8 pitch lengths long).

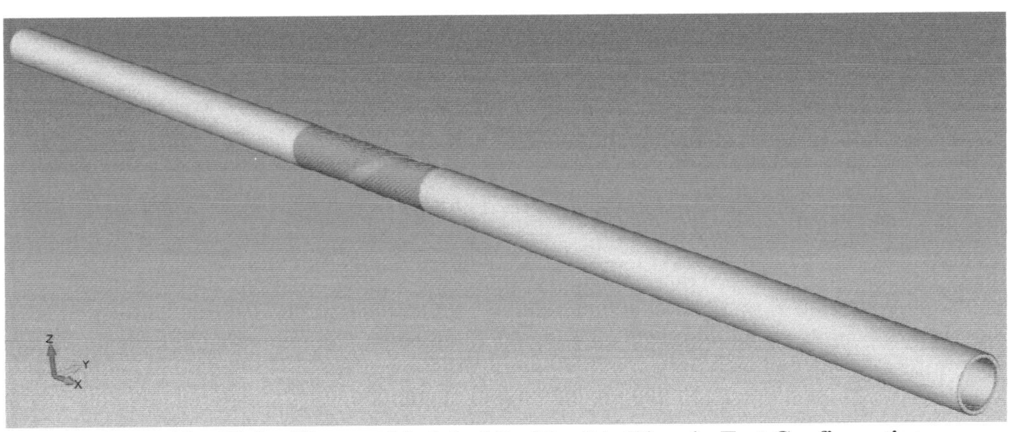

Figure 15 – Finite Element Model of the Flexible Riser in Test Configuration

Figure 16 provides the comparison of wire strains for all 34 outer wires for the 3 broken wire configuration.

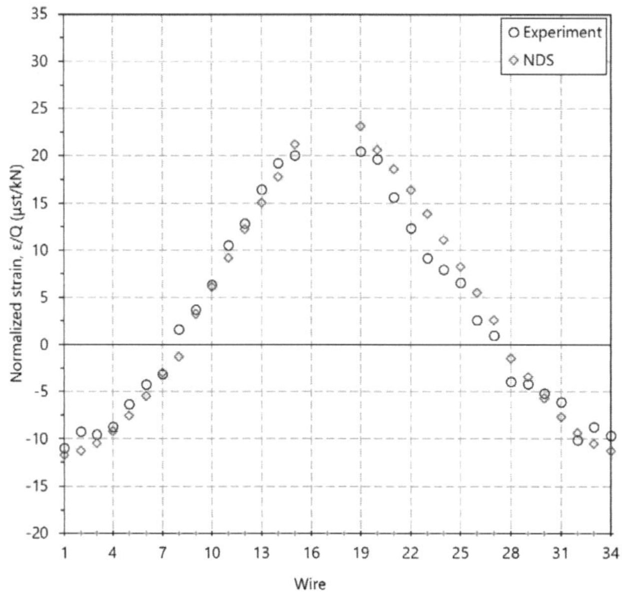

Figure 16 – Comparison of Wire Strains between Experiment and Finite Element Model with 3 Broken Wires

It is important to note that the computational speed of the accelerated solver was not impacted due to the more onerous broken wire kinematics – for these cases, the computation was executed in nearly real-time.

SENSORS AND INSTRUMENTATION

This paper is focused on flexible riser digital twin for the hang-off section – the dynamic riser. Here the direct use of sensor rotational acceleration measurements as input to the FEM in the analysis configuration (Figure 7) is demonstrated. This is compared against the typical rotational displacement input. It is noteworthy to distinguish each implementation so that it is clear that the same results should be expected. The fact that the 6 DoF accelerometer measurements are readily available from riser and vessel mounted inertial measurement units tilts the balance towards the direct implementation of rotational acceleration inputs to the FEM. Simply stated, rotational displacements are enforced via the stiffness matrix whereas rotational acceleration inputs are enforced via the mass matrix. This should not be surprising since the accelerations and displacements are kinematic quantities related by the numerical integration process.

For this demonstration, 4 degrees of top rotation at a period of 1 seconds for 10 seconds is executed along with a tension of 1000 kN (constant, assume measured via hang-off force transducer). Cross section moment time-histories are extracted 7 pitches down from the hang-off for both displacement (Figure 17) and acceleration (Figure 18) driven analysis configurations. As it can be seen, the results are identical.

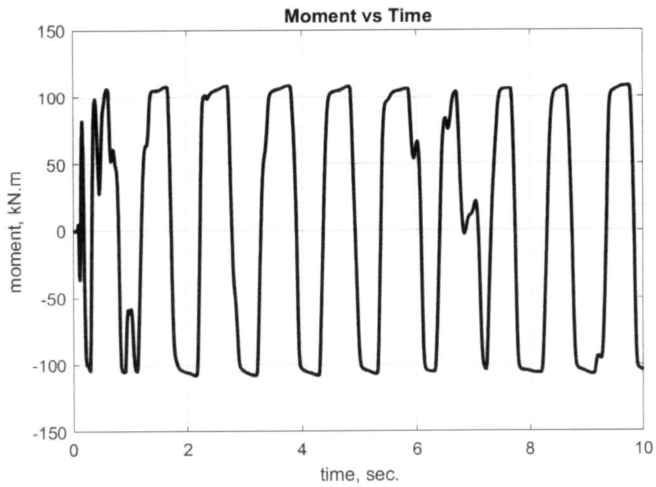

Figure 17 – Cross Section Moment Time-History (7 Pitch Lengths from Hang-off) for Displacement Driven Rotational Inputs

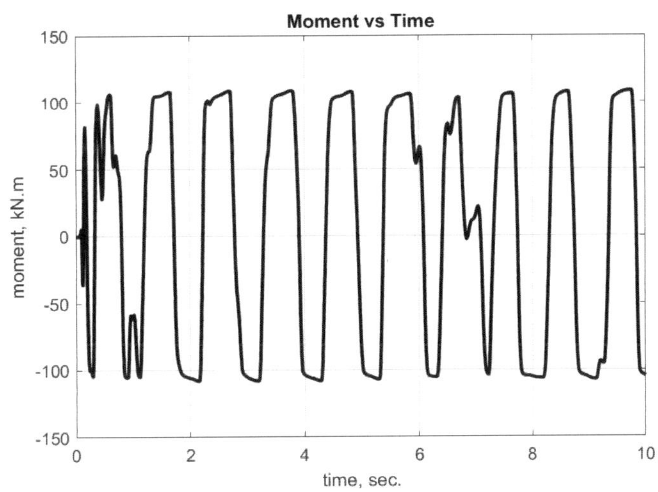

Figure 18 – Cross Section Moment Time-History (7 Pitch Lengths from Hang-off) for Acceleration Driven Rotational Inputs

APPROXIMATE METHODS

The high-fidelity 3D FEM and accelerated solver combination constitute a physics-based optimal solution for real-time resolution of wire stresses and remnant fatigue life calculations. Prior to the introduction of accelerated solvers, the method for resolving irregular wave wire stresses involved interpolation and post-processing of riser curvature and tension data. In such an approach, wire stress hysteresis (due to 1 cycle of harmonic input) are generated for a family of tension and pressure combinations. These "regular wave" hysteretic functions are then utilized as a data base of "table look-ups" to construct the irregular wave wire stress. The issue with such an approximation is that starting with the 2nd cycle of bending, some armor wires may initiate their transition from stick to slip from a different location than the first cycle regular wave hysteresis curves. As such, the basis for such an irregular wave approximation may quickly become insufficient as the cycles progress and the complex features of stick/slip friction physics, such as wire ratchetting behavior, become dominant.

Another approximate approach for flexible riser wire stress calculations could involve "machine-learning" or "neural-networks". In such an approach, simulation-based methods such as high-fidelity 3D FEM and accelerated solvers are utilized to first solve a large number of load cases and extract wire stresses at the 4 corners of each wire at each critical cross section. This data is then utilized to "train" the machine. The drawbacks in such an approach include difficulties in producing the training data sets (accelerated solvers change this equation) and the uncertainty in the predictive accuracy especially in light of subject problem complexities. In reality, a single wire slipping a bit earlier due to a slight shift in inputs can completely change the remaining wire stress time-history in a nonlinear problem.

FLEXAS FIELD APPLICATION – SHELL BONGA FLEXIBLES LIFE EXTENSION PROJECT

The Shell/Bonga flexibles life extension project involved inspection and integrity assessment of two 18.75" flexible offloading risers. The inspections of the tensile armor wires were executed by a partner company utilizing a magnetic eddy current scanner to detect metal loss, cracks, and outsized wire gaps. Both the type and severity of the defects were detected.

The integrity assessment involved (1) building a global model, (2) building a local model inclusive of any damage detected by the inspection, (3) executing the irregular wave fatigue simulations (FLEXAS accelerated solver), and (4) post-processing the wire stress time-histories (at each wire's corners) to generated irregular wave fatigue damage spectra and compute remnant fatigue life.

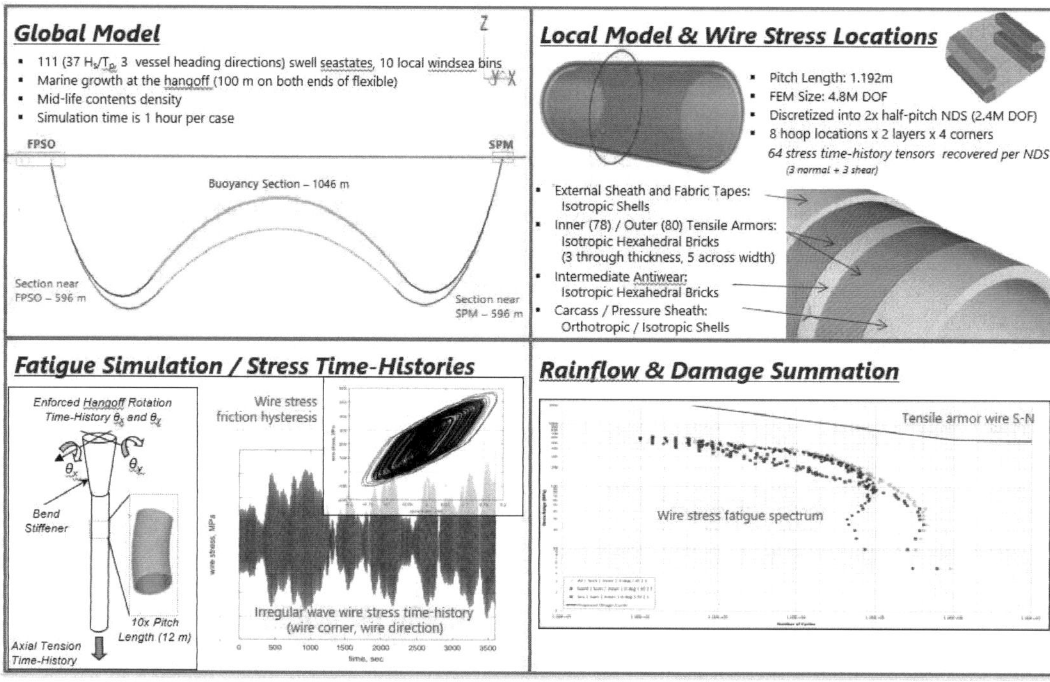

Figure 19 – Shell/Bonga Flexibles Irregular Wave Fatigue Simulations Process

The successful execution of this project required an accelerated solver capability to execute 1000s of irregular wave bending cycles, essentially in real-time, on a high-fidelity flexible riser local FEM inclusive of damaged wires (if detected by inspection). It should be intuitive that flexible riser life extension must involve precise and accurate capture of wire kinematics and the irregular wave environments. Simplified flexible riser representations (e.g., Figure 3, left figure), simplification of dynamic environments from irregular to regular (Figure 13), and modeling damage with Stress Concentration Factors instead of the actual damage (Figure 14) would likely introduce enough analysis errors to render conclusions relative to flexible riser life extension inconclusive. In addition, it should be noted here that although it is well recognized that simplifications/idealizations impact accuracy and predictive capabilities, it would be incorrect to assume that they would automatically result in a more conservative result. The degree of conservativism, if any, can only be gaged if the more precise solution is first available. Also, for flexible riser integrity management applications, any rational approach must involve high precision and accuracy as presented here. As a final note, it should be clear from this section that a flexible riser real-time structural integrity management solution based on an accelerated solver has been established and that bridging this advanced capability to support a digital twin platform is feasible and practical.

CONCLUSION

High-fidelity FEMs and accelerated solvers are essential components for a flexible riser real-time structural integrity management solution. The accelerated solver is continuously resolving, in real-time, wire stresses at critical cross sections driven by direct sensor data composed of riser hang-off rotational accelerations and tension. These wire stress time-histories are continuously post-processed for irregular wave fatigue spectra and remnant life. An initial inspection of the riser is carried to detect existing damage to the wires which is also incorporated into the high-fidelity FEM. It was demonstrated that high-fidelity 3D FEM can capture damaged wire kinematics and therefore the impact the remnant fatigue life.

Flexible riser structural digital twins can provide real-time actionable data to the operator for remnant fatigue life monitoring as well as extreme events. This provides guidance for when and where to inspect on the flexible riser – with the locations of interest based on predicted fatigue damage accumulation. Such guidance may prevent catastrophic failure of the flexible riser and the associated costs of replacement and production loss.

ACKNOWLEDGEMENTS

A word of thanks to ExxonMobil, INTECSEA, and Worley Digital for supporting this work.

REFERENCES

Perdrizet, T., Leroy J., Barbin, N., Le-Corre, V., Charliac, D., and Estrier, P., "Stresses in Armour Layers of Flexible Pipes: Comparison of Abaqus Models", 2011 SIMULIA Customer Conference.

Majed, A, Rombado, G, Chinello, L, Cooke, N and Kan, W. "Computationally Efficient Simulation of Flexible Risers via Nonlinear Dynamic Substructures: Numerical and Experimental Validation", *Proc 26th Int Offshore and Polar Eng Conf*, Rhodes, Greece, ISOPE, 2016.

Majed, A, Yiu, F, Cooke, N, Chinello, L, Gomes, J and Kusinski, G. "DeepStar Global Offshore Technology Development Program: New Generation Computational Capabilities in Nonlinear Dynamic Simulations of Flexible Riser Systems", *Proc Offshore Technology Conference*, Houston, Texas, USA, OTC, 2017.

Majed, A, Chinello, L, Cooke, N, Ni, X. "A Simulation-Based Approach to Irregular Wave Wire Stress Computations in Flexible Risers", *Proc 27th Int Offshore and Polar Eng Conf*, San Francisco, USA, ISOPE, 2017.

Majed, A., Chinello, L., and Cooper, P., "Computationally Efficient Flexible Pipe Bench Test Simulations Using Nonlinear Dynamic Substructures", Proceedings of International Society of Offshore and Polar Engineers (ISOPE), Kona, Hawaii. 2015.

Doynov, K., Rombado, G., Cooke, N., and Majed, A., "Validation of Predictions of Wire Stress of Flexible Pipe with Damaged Tensile Armor Wires Under Combined Tension and Bending", Proceedings of ASME 2020 International Conference on Ocean, Offshore and Arctic Engineering (OMAE2020), Fort Lauderdale, FL, 2020.

deSousa, J, R., M, Magluta, C, Roitman, N, Campello, "Bending Analysis of a Flexible Pipe with Broken Tensile Armor Wires", *Proc 34th Int Conf on Ocean, Offshore and Arctic Engineering*, St. John's, Canada, OMAE 2015.

IDENTIFYING KEY SAFETY CULTURE FACTORS FOR THE OFFSHORE INDUSTRY

NING LOU
RESEARCH ASSISTANT, UNIVERSITY OF HOUSTON

EZRA WARI
RESEARCH ASSISTANT, UNIVERSITY OF HOUSTON

JAMES CURRY
ASSOCIATE PROFESSOR, LAMAR UNIVERSITY

KEVIN MCSWEENEY
MANAGER, CORPORATE TECHNOLOGY - HUMAN FACTORS, SAFETY, AND EMERGING INSPECTION TECHNOLOGY, AMERICAN BUREAU OF SHIPPING

RICK CURTIS
PRINCIPLE RISK AND SAFETY CULTURE ENGINEER, ABS GROUP

BRIAN CRAIG
PROFESSOR AND DEAN OF ENGINEERING, LAMAR UNIVERSITY

MUHAMMAD MUZAMIL HUSSAIN
RESEARCH ASSISTANT, UNIVERSITY OF HOUSTON

WEIHANG ZHU
ASSOCIATE PROFESSOR, UNIVERSITY OF HOUSTON

Proceedings of the 26th Offshore Symposium, April 6th, 2021, Houston, Texas
Texas Section of the Society of Naval Architects and Marine Engineers

ABSTRACT

This research identifies key factors, or safety culture categories, that can be used to help describe the safety culture for the offshore oil and gas industry and develop a comprehensive offshore safety culture assessment toolkit for use by the US Gulf of Mexico (GoM) owners and operators. Detailed questionnaires from selected safety culture frameworks of different industries were collected and analyzed to identify important safety culture factors and key questions for assessment. Safety frameworks from different associations were investigated, including the Center for Offshore Safety (COS), Bureau of Safety and Environmental Enforcement (BSEE), and the National Transportation Safety Board (NTSB). The safety culture factors of each of these frameworks were generalized and analyzed. The frequency of the safety culture factors in each framework was analyzed to explore commonality. The literature review and analysis identified a list of common factors among safety culture frameworks.

Keywords: Safety Culture, Literature Review, Risk, Human Factors, Oil and Gas Industry

1. INTRODUCTION

Offshore oil and gas production is an inherently hazardous activity that could potentially fail in complex, catastrophic ways. The offshore sector is generally high-risk due to the harsh environments, hazardous materials handled, congested equipment arrangements, and remote locations with unique transportation issues (Broni-Bediako, 2010). A significant example is the accident which happened ten years ago with the Deepwater Horizon in the Gulf of Mexico followed by an 87-day-long of oil spill, causing serious consequences immediately to the present day (Skogdalen, 2010). While many factors led to this disaster, several reports issued in the aftermath emphasized the need for an improved safety culture in the offshore industry. It has been suggested that a robust safety culture has many dimensions, including leadership commitment to safety values and actions, a safe environment for raising concerns or reporting incidents and near misses, and organizational policies and practices that encourage employees to make safe decisions (NASEM, 2021).

There is a general perception in the offshore industry that more rules, regulations, and procedures are unlikely to improve safety performance. Instead, the industry needs a better understanding of the social and organizational factors that foster professionalism during routine and emergency situations. The impact of safety culture on the outcome of positive safety performance is well recognized and emphasizes that operational incidents are not just a result of human error, technical failures, or environmental factors alone, but a combination of those and others such as failure of management, employees not performing their duties, and a breakdown in documented systems. A mature safety culture helps overall operational and safety performance. There should be a close relationship between an organization's safety culture and Safety Management Systems (SMS). An effective SMS must take account of all factors that have an impact on safety including human and organizational needs. Conversely, the safety culture influences the way in which the SMS is integrated and implemented within the organizational culture. Consequently, the assessment of safety culture and the SMS should be complementary.

The goal of this project is to develop a roadmap that the industry can use to evaluate and improve organizational safety culture, reduce unsafe behaviors, improve individual performance, and reduce management system failures. The final deliverables will include a safety culture evaluation toolbox and data analysis methods to identify what actions have been or could be successful in improving safety. This paper reports the project's progress to date.

Our research seeks to identify key factors, or safety culture categories, that can be used to describe the safety culture for the offshore oil and gas industry to aid in the development of an offshore safety culture assessment toolkit for use by the US Gulf of Mexico owners and operators. This research involves a thorough benchmarking of available safety culture assessment methodologies, measurement frameworks, quantitative and qualitative data collection methods, statistical evaluation approaches, and means for improving safety culture. The research started with a literature review using keywords such as "safety", "culture", "climate", "offshore", "maritime", and "oil and gas".

The scope of the paper discusses a multi-year project intended to learn from past industry activities related to evaluating and understanding safety culture. Key portions of the project's first year activities are the core of this paper. This paper is organized as follows. Section 2 defines the scope of the literature survey and compares two relevant concepts, "safety culture" and "safety climate"; Section 3 provides the literature survey of safety culture and safety climate; Section 4 discusses the frequency of occurrence of safety culture factors in questionnaires; the paper is concluded in Section 5. The conclusion also includes an overview of future project activities.

2. Defining Safety Culture and Safety Climate

2.1 Defining safety culture

Safety culture is a critical part of the overall culture of an organization. The term first became popular following the Chernobyl nuclear disaster when it was suggested that organizations can reduce accidents and safety incidents by developing a "positive safety culture" (INSAG, 1992). The U.S. Nuclear Regulatory Commission defines safety culture as "the core values and behaviors resulting from a collective commitment by leaders and individuals to emphasize safety over competing goals to ensure protection of people and the environment." (National Academies of Sciences, 2016) The Bureau of Safety and Environmental Enforcement (BSEE), a regulator of the offshore oil and gas industry, issued its *Safety Culture Policy Statement* in May 2013 to promote safety culture in the industry. It defined safety culture as "the core values and behaviors of all members of an organization that reflect a commitment to conduct business in a manner that protects people and the environment" and articulated nine characteristics or elements of a robust safety culture:

1) Leadership commitment to safety values and actions,
2) Respectful work environment,
3) Environment for raising concerns,
4) Effective safety and environmental communication,
5) Personal accountability,
6) Inquiring attitude,
7) Hazard identification and risk management,
8) Work processes, and
9) Continuous improvement.

Although there are inconsistencies with the numerous definitions of safety culture that have been proposed in the literature, there is considerable overlap in the key aspects of those definitions. Those articulated by BSEE align with those identified in major scholarly reviews of safety culture research and leading definitions in other industries and are grounded in empirical research. Table 1 lists some selected definitions of safety culture that have been proposed by different researchers and institutes.

Table 1 Selected definitions of Safety Culture

Definition according to the publication	**(Author, year)**
Safety culture – established convictions, norms, attitudes, roles and technical procedures referring to minimizing the exposition of employees, managers, clients and members of the society to conditions regarded as dangerous or harmful.	(Turner, 1989, November)
Safety cultures reflect the attitudes, beliefs, perceptions, and values that employees share in relation to safety.	(Cox S. C., 1991)
Culture is conceptualized as primarily an ideational system of meanings, and safety culture as one concerned with the norms, beliefs, roles, and practices for handling hazards and risk.	(Pidgeon N. F., 1991)
The safety culture of an organization is the product of individual and group values, attitudes, competencies, and patterns of behavior that determine the commitment to, and the style and proficiency of, an organization's health and safety management.	(Health and Safety Commission, 1993)
A safety culture is the set of assumptions, and their associated practices, which permit beliefs about danger and safety to be constructed.	(Pidgeon N. , 1997)
An abstract concept, which is underpinned by the amalgamation of individual and group perceptions, thought processes, feelings and behaviors, which in turn gives rise to the particular way of doing things in the organization. It is a sub-element of the overall organizational culture.	(Kennedy, 1998)
Safety culture' may be perceived as a subset of organizational culture, where the beliefs and values refer specifically to matters of health and safety.	(Clarke, 1999)
The product of multiple goal-directed interactions between people (psychological), jobs (behavioral) and the organization (situational); while safety culture is that observable degree of effort by which all organizational members directs their attention and actions toward improving safety on a daily basis.	(Cooper M. D., 2000)
Comprises attitudes, behaviors, norms and values, personal responsibilities as well as human resources features such as training and development.	(Glendon, 2000)
Those aspects of the organizational culture which will impact on attitudes and behavior related to increasing or decreasing risk.	(Guldenmund F. W., 2000)
Safety culture is "the attitudes, beliefs and perceptions" shared by natural groups as defining norms and values", which determine how they act and react in relation to risks and risk control systems.	(Hale, 2000)
A sub-facet of organizational culture, which affects workers' attitudes and behavior in relation to an organization's on-going safety performance.	(Mohamed, 2003)
Shared and learned meanings, experiences and interpretations of work and safety - expressed partially symbolically – which guide people's actions towards risk, accidents, and prevention.	(Richter, 2004)
A set of prevailing indicators, beliefs, and values that the organization owns in safety.	(Fang, 2006)
A dynamically balanced, adaptable state resulting from the configuration of values, leadership strategies, and attitudes that collectively impact safety performance at the individual, group, and enterprise level. Simply stated, safety culture is a dynamic configuration of factors at multiple levels that influences safety performance.	(Patankar, 2010)
Safety culture can be viewed as the assembly of underlying assumptions, beliefs, values, and attitudes shared by members of an organization, which interact with an organization's structures and systems and the broader contextual setting to result in those external, readily visible, practices that influence safety.	(Edwards, 2013)

2.2 *DEFINING SAFETY CLIMATE*

Throughout the review, the terms of "safety culture" and/or "safety climate" have been mentioned by almost all the relevant papers. Cox and Flin (1998) commented that the terms safety culture and safety climate are often used interchangeably to refer to similar concepts. Culture, as a topic, is viewed as much harder to study and measure (Barrett, 1981), while climate is considered to be more directly observed and can provide an indication of people's perception and cognitive interpretation of organizational practices (Schein, 1990). The difference may also come from how one tries to assess each of the two terms. "Climate" has usually been assessed using quantitative methods, while "culture" is generally examined with qualitative methods. Some definitions by different researchers and institutes are selected into the Table 2.

Table 2 Selected definitions of Safety Climate

Definition according to the publication	Author/Year
Safety climate reflects employees' perceptions about the relative importance of safe conduct in their occupational behavior. It can vary from highly positive to a neutral level, and its average level reflects the safety climate in a given company.	(Zohar, 1980)
Safety climate refers to a set of attributes that can be perceived about particular work organizations (maintenance, construction, and central repair shops) and which may be induced by the policies and practices that those organizations impose upon their workers and supervisors.	(Niskanen, 1994)
The objective measurement of attitudes and perceptions toward occupational health and safety issues.	(Coyle, 1995)
Safety climate is concerned with the shared perceptions and beliefs that employees hold regarding safety in their workplace. These serve to construct a self-sustaining image of risk, danger and safety in an organization.	(Cooper D. , 1997)
[Safety] Climate is defined as a set of molar perceptions, shared by individuals with their work environment, which are valid as references for guiding behavior in the execution of tasks during day-to-day eventualities.	(Isla Díaz, 1997)
Climate can be viewed as a temporal state measure of culture, which is reflected in the shared perceptions of the organization at a discrete point in time.	(Cheyne, 1998)
Safety climate is a specific form of organizational climate, which describes individual perceptions of the value of safety in the work environment.	(Neal, 2000)
Safety climate should be conceptualized as a higher order factor comprised of more specific first-order factors. The first-order factors of safety climate should reflect perceptions of safety-related policies, procedures, and rewards. The higher order factor of safety climate should reflect the extent to which employees believe that safety is valued within the organization.	(Griffin, 2000)
Safety climate is the temporal state measure of safety culture, subject to commonalities among individual perceptions of the organization. It is therefore situationally based, refers to the perceived state of safety at a place at a particular time, is relatively unstable, and subject to change depending on the features of the current environment or prevailing conditions.	(Wiegmann, 2002)
The concept of safety climate refers to employees' perceptions of the relative emphasis placed by management on safety issues relative to other organizational concerns.	(Fogarty, 2010)
Safety climate refers to how employees perceive the enactment of organizational policies and procedures relating to safety in their organization at a given point in time.	(Mearns, 2013)

Despite some inexplicit differences, the commonalities are overwhelming between safety culture and safety climate. This may cause confusion in definitions and the relationship between them. However, it should be noted that these two terms are not the same thing when trying to assess either of them. (Denison, 1996) (Guldenmund F. W., 2010) While we recognize the differences between safety culture and safety climate, we will use them interchangeably in this paper for simplicity.

3. Results of the Safety Culture/Climate Assessment Tools Literature Review

Safety culture and safety climate are difficult to model if they are abstract and conceptual. Research has previously been conducted to create a means to evaluate them, either quantitatively or qualitatively. We reviewed assessment processes that are considered to be efficient and accurate. We collected 21 frameworks. Among them, 16 safety culture assessment methodologies, or tools, were studied in more details which cover many industry sectors including offshore, aviation, construction, and railway. A summary list can be found in The tool is a paper-based questionnaire consisting of 153 statements, that either require true/false answers or answers on a five-point Likert-type scale. There were three versions developed by Robert Gordon University and Aberdeen University. The questionnaire collects information on the following:

- General information
- Job (work pressure, work clarity, job communication, safety behavior)
- Risk perception - main hazards
- Risk perception - work task hazards
- Assessment of safety
- Safety attitudes
- Contractor safety attitudes* *15 statements, only to be answered if employed by a contracting company
- Job security
- Accident history

The 'Risk perception - main hazards' statements cover 3 factors:

- Hazards to the installation
- Occupational hazards
- Catastrophes

The 'Assessment of safety' statements cover 3 factors:

- Accident prevention
- Incident mitigation
- Emergency response

The 'Safety attitudes' statements cover 10 factors:

- Speaking up about safety
- Attitude to violations
- Supervisor commitment to safety
- Attitude to rules and regulations
- OIM commitment to safety
- Safety regulation
- Cost versus safety
- Personal responsibility for safety

- Safety systems
- Over-confidence in own safety

The 'Contractor safety attitudes' statements cover 3 factors:

- Management commitment to safety
- Contractor company commitment to safety
- Confidence in operating company

Health and Safety Climate Survey Tool (HSCST)

The Health and Safety Climate Survey Tool (HSCST) was developed by the U.K. Health & Safety Executive (HSE) and also released in December 1997. It is a generically designed questionnaire without any O&G industry specific items. With proper customization, it could be used in other industries.

As surveyed, most safety tools are developed and designed in hierarchical structures. While focusing on certain areas, they are built upon multiple "safety factors" as detailed "dimensions". For instance, LSCAT covers four broad areas with nine dimensions, or "factors", as discussed above. The HSCST covers the following ten factors:

- Organizational commitment and communication
- Line management commitment
- Supervisors' role
- Personal role
- Workmate's influence
- Competence
- Risk taking behavior and some contributory influences
- Some obstacles to safe behavior
- Permit-to-work systems
- Reporting of accidents and near misses

Most safety culture assessment frameworks cover multiple dimensions or factors that often appear to be similar. The process safety culture chapter of *Guidelines for Risk Based Process Safety* covers the most complete list of safety culture factors summarized as follows:

1) Core Value: Establish safety as a core value
2) Strong Leadership: Provide strong leadership
3) Accountability: Establish and enforce high standards of performance
4) Effective Culture Management System (MS): Document the safety culture emphasis/approach
5) Sense of Vulnerability: Maintain a sense of vulnerability
6) Empowerment: Empower individuals to successfully fulfill their safety responsibilities
7) Defer to Expertise: Maintain the "critical mass" of expertise required for safe operations
8) Effective Communications: Ensure open and effective communications
9) Safe Questioning/Learning Environment: Establish a questioning/learning environment
10) Mutual Trust: Foster mutual trust
11) Responsiveness: Provide timely response to safety issues and concerns
12) Continuous Monitoring: Provide continuous monitoring of performance

Table 3.

The assessment tools generally have multiple groups in the questionnaire structure (also known as safety culture "factors" or "dimensions"). Within each group, there are usually several questions to captures the main characteristics of that group. Of the 16 methodologies reviewed in this study, there were six safety culture tools that were developed or can be adapted for the Oil and Gas (O&G) industry. Three of them that can be used for offshore industry are highlighted below.

Loughborough University Safety Climate Assessment Toolkit (LSCAT)

The Loughborough University Safety Climate Assessment Toolkit (LSCAT) was developed and used in a joint industry project "The measurement of safety climate in safety cases" in 1997. LSCAT is open to the public and designed to be used without expert external assistance, although some companies may need some support. Services available from the developer include providing a mean score on each dimension based on existing databases, safety climate profiles, benchmarking, and a comprehensive user guide in Q&A format. It covers four dimensions (broad areas):

- Organizational context
- Social environment
- Individual appreciation
- Work environment

A brief description of each dimension from the LSCAT is given below:

Organizational Context ·

- Management Commitment - Perceptions of management's overt commitment to health and safety issues.
- Communication - The nature and efficiency of health and safety communications within the organization.
- Priority of Safety - The relative status of health and safety issues within the organization.
- Safety Rules and Procedures - Views on the efficacy and necessity of rules and procedures.

Social Environment

- Supportive Environment - The nature of the social environment at work, and the support derived from it.
- Involvement - The extent to which safety is a focus for everyone and all are involved.

Individual Appreciation

- Personal Priorities and Need for Safety - The individual's view of their own health and safety management and need to feel safe.
- Personal Appreciation of Risk - How individuals view the risk associated with work

Work Environment

- Physical Work Environment - Perceptions of the nature of the physical environment

Also, there may be "Organization Specific Factors" which are the attitudes to specific safety related systems and procedures (for example, permit to work systems) may be included, as necessary.

LSCAT is best suited for those with some background in safety climate assessment. Advice is given on the feedback and climate improvement processes, including brief details on developing action plans and a safety climate maintenance checklist. It was developed for use in the offshore O&G industry but can be adapted to other sectors, while subject to industry-specific modifications.

Offshore Safety Questionnaire (OSQv1)

The tool is a paper-based questionnaire consisting of 153 statements, that either require true/false answers or answers on a five-point Likert-type scale. There were three versions developed by Robert Gordon University and Aberdeen University. The questionnaire collects information on the following:

- General information
- Job (work pressure, work clarity, job communication, safety behavior)
- Risk perception - main hazards
- Risk perception - work task hazards
- Assessment of safety
- Safety attitudes
- Contractor safety attitudes* *15 statements, only to be answered if employed by a contracting company
- Job security
- Accident history

The 'Risk perception - main hazards' statements cover 3 factors:

- Hazards to the installation
- Occupational hazards
- Catastrophes

The 'Assessment of safety' statements cover 3 factors:

- Accident prevention
- Incident mitigation
- Emergency response

The 'Safety attitudes' statements cover 10 factors:

- Speaking up about safety
- Attitude to violations
- Supervisor commitment to safety
- Attitude to rules and regulations
- OIM commitment to safety
- Safety regulation
- Cost versus safety
- Personal responsibility for safety
- Safety systems
- Over-confidence in own safety

The 'Contractor safety attitudes' statements cover 3 factors:

- Management commitment to safety
- Contractor company commitment to safety
- Confidence in operating company

Health and Safety Climate Survey Tool (HSCST)

The Health and Safety Climate Survey Tool (HSCST) was developed by the U.K. Health & Safety Executive (HSE) and also released in December 1997. It is a generically designed questionnaire without any O&G industry specific items. With proper customization, it could be used in other industries.

As surveyed, most safety tools are developed and designed in hierarchical structures. While focusing on certain areas, they are built upon multiple "safety factors" as detailed "dimensions". For instance, LSCAT covers four broad areas with nine dimensions, or "factors", as discussed above. The HSCST covers the following ten factors:

- Organizational commitment and communication
- Line management commitment
- Supervisors' role
- Personal role
- Workmate's influence
- Competence
- Risk taking behavior and some contributory influences
- Some obstacles to safe behavior
- Permit-to-work systems
- Reporting of accidents and near misses

Most safety culture assessment frameworks cover multiple dimensions or factors that often appear to be similar. The process safety culture chapter of *Guidelines for Risk Based Process Safety* (Center for Chemical Process Safety, 2007) covers the most complete list of safety culture factors summarized as follows:

13) Core Value: Establish safety as a core value
14) Strong Leadership: Provide strong leadership
15) Accountability: Establish and enforce high standards of performance
16) Effective Culture Management System (MS): Document the safety culture emphasis/approach
17) Sense of Vulnerability: Maintain a sense of vulnerability
18) Empowerment: Empower individuals to successfully fulfill their safety responsibilities
19) Defer to Expertise: Maintain the "critical mass" of expertise required for safe operations
20) Effective Communications: Ensure open and effective communications
21) Safe Questioning/Learning Environment: Establish a questioning/learning environment
22) Mutual Trust: Foster mutual trust
23) Responsiveness: Provide timely response to safety issues and concerns
24) Continuous Monitoring: Provide continuous monitoring of performance

Table 3 Selected safety culture/climate assessment tools

Title/Name (Acronym)	**Developer/Author**	**Country of origin**	**Sector/Industry**
Loughborough Safety Climate Assessment Toolkit (LSCAT)	Loughborough University, Health & Safety Executive (HSE), and some offshore organizations	UK	Offshore O&G, broad use
Safety Health of Maintenance Engineering (ShoMe) Tool	UK Civil Aviation Authority (CAA), developed by Health and Safety Engineering Consultants (HSEC) Ltd.	UK	Aviation
Safety Culture Toolbox	Eurocontrol, Aberdeen University	EU	Air Navigation Service Providers (ANSP)

Her Majesty's Railway Inspectorate (HMRI) Safety Culture Inspection Toolkit	Human Engineering Ltd.	UK	Railway
RSSB Safety Culture Improvement Toolkit	Rail Safety and Standards Board (RSSB)	UK	Railway
Multilevel Safety Climate (MSC) Scale	Zohar (1980), Zohar and Luria (2005)	Israel	Manufacturing
ABS Guidance Notes on Safety Culture and Leading Indicators of Safety	American Bureau of Shipping (ABS)	USA	Maritime
Commercial Aviation Safety Survey (CASS)	Federal Aviation Administration (FAA), University of Illinois	USA	Aviation
Norwegian Offshore Risk and Safety Climate Inventory (NORSCI)	International Research Institute of Stavanger (IRIS)	Norway	Offshore
Nordic Occupational Safety Climate Questionnaire (NOSACQ)	Consortium of Scandinavian organizations	Nordic countries	Construction/high-risk industries
Offshore Safety Questionnaire (OSQv1, OSQ99, CSCQ)	Robert Gordon University, Aberdeen University and HSE	UK	O&G
Health and Safety Climate Survey Tool (HSCST)	Health & Safety Executive (HSE)	UK	Initially for O&G but can be used in any other industry with proper customization.
Quest Safety Climate Questionnaire (QSCQ)	Quest Evaluations and Databases Ltd	UK	O&G industry
Guidelines for Risk Based Process Safety	Center for Chemical Process Safety	USA	Chemical Process Industry

The following table summarizes the number of safety culture factors covered by each framework. Please see Appendix I for the abbreviations. It can be seen that different frameworks vary significantly in terms of the number of safety culture factors.

Table 4 Number of Safety Culture Factors Covered by Each Framework

Framework	No. of factors	Framework	No. of factors
CCPS RBPS	12	COS	6
CCPS Core	10	CNEB	6
PHMSA	10	HRO	5
BSEE	9	Reasons	5
ABS	8	PAME	5
Japan OGI	8	NTSB	5
NOPSEMA	8	CCHS	4
UK HSE	7	UK PSLG	4

INPO	7	IOGP	4
NRC	7	IAEA	4

Table 5 displays the number of frameworks, or tools, that include a specific safety culture factor. When individual factors are examined, not one of the factors is included in all frameworks. For instance, most factors have leadership as a separate safety culture factor but 4 of the 12 did not have a separate leadership factor. While all frameworks discussed leadership, several did not make it an independent factor. Only 5 factors were common to over half the frameworks. The organization of factors was surprisingly inconsistent given the frameworks are describing the common subject of safety culture.

Table 5 Number of frameworks covering each factor out of 21 frameworks.

Factor	Covered by Number of frameworks
Strong Leadership	17
Safe Questioning/Learning Environment	15
Empowerment	13
Effective Communications	13
Continuous Monitoring	12
Sense of Vulnerability	11
Core Value	10
Mutual Trust	9
Accountability	8
Effective Culture MS	7
Defer to Expertise	3
Responsiveness	3

4. Safety Factor Keywords for an Offshore Safety Culture Framework

It is noted that the frameworks with fewer factors covered are not inferior or under-developed frameworks. They are simply based on different focus areas. From what we learned from the literature review, a comprehensive framework needs to consider a hierarchical structure that covers four basic areas, similar to those covered by the LSCAT toolkit. They are:

A. Organizational level
B. Management performance
C. Personnel/workforce
D. Environment outside the organization

Figure 1 illustrates the framework structure. The management level is built upon the workforce level. Meanwhile, the subject encompassing these two is the organizational level, which acts as the structural factor that connects and regulates the management and workforce levels. On the outside, there is the environmental factor that represents the overall external regulatory force, such as governmental acts, that usually has a dominant impact upon the safety culture of an organization.

Our framework is built upon the one developed from Center for Chemical Process Safety as well as Bureau of Safety and Environmental Enforcement (BSEE) safety culture policy. The critical task here is to identify the suitable safety culture factors, which act like the "bricks", for this building.

Figure 1 The structure of our proposed safety framework

One way to find out which safety culture factors are the best fit for a given application is to review perception survey questions from the literature. As summarized in Table 6, these questionnaires typically have different factor-driven focus areas and specifications. These questionnaires may come from publications, safety toolkits, manuals, or booklets.

Typically, the survey questions are aligned, or mapped, to specific safety culture factors. That was done in this study with the perception survey questions aligned to all the safety culture factors listed in Table 6. Many of these are very similar. A key word search was conducted on each of the available survey question sets which allowed us to group the questions by main content and subsequently map them to safety culture factors. Table 7 contains the results of the keyword search noting the count of number of occurrences that the keyword appeared in the search.

Table 6 Questionnaires and Safety Factors from the Literature

Title	Country	Industry	# of safety factors
Oltedal, H. A., & McArthur, D. P. (2011)	Norway	Maritime	7
Oltedal, H. A., & Engen, O. A. (2009)	Norway	Maritime	5
Hystad, S. W., Saus, E. R., Sætrevik, B., & Eid, J. (2013)	Norway	Offshore	7
Shirali, G. A., Shekari, M., & Angali, K. A. (2016)	Iran	Oil& Gas	13
Cox, S. J., & Cheyne, A. J. T. (2000)	UK	Offshore	9
Stemn, E., Bofinger, C., Cliff, D., & Hassall, M. E. (2018)	Ghana	Mining	13
Morrow, S. L., Koves, G. K., & Barnes, V. E. (2014)	USA	Nuclear	9
Reiman, T., & Pietikäinen, E. (2012)	Finland	General	12
Tang, D. K. H., Leiliabadi, F., & Olugu, E. U. (2017)	Malaysia	Oil & Gas	14
Nielsen, M. B., Eid, J., Hystad, S. W., Sætrevik, B., & Saus, E. R. (2013)	Norway	Maritime	7
Hon, C. K., Chan, A. P., & Yam, M. C. (2014)	Hong Kong	Construction	5
Håvold, J. I., Ghulam, M., & Ashraf, N. (2017, September)	Norway	Maritime	10
Zahoor, H., Chan, A. P., Utama, W. P., Gao, R., & Zafar, I. (2017)	Pakistan	Construction	7
Zahoor, H., Chan, A. P., Utama, W. P., Gao, R., & Memon, S. A. (2017)	Pakistan	Construction	5
Curcuruto, M., Griffin, M. A., Kandola, R., & Morgan, J. I. (2018)	UK	Railway	14
Coyle, I. R., Sleeman, S. D., & Adams, N. (Coyle, 1995)	Australia	Service	8
National Academies of Sciences, Engineering, and Medicine. (2016)	USA	Offshore	9
Høivik, D., Moen, B. E., Mearns, K., & Haukelid, K. (2009)	Norway	Oil & Gas	7
Grabowski, M., You, Z., Song, H., Wang, H., & Merrick, J. R. (2009)	USA	Maritime	18
Oltedal, H. A. (2011)	Norway	Maritime	11
Hearts and Minds Toolkit: Learning from incidents questionnaire (2004)	UK	Oil & Gas	5

Table 7 Safety culture factor keywords and frequency from the literature

Keyword	Count	Keyword	Count
Accountability	9	Identification	3
Awareness	2	Informed	2
Climate	0	Inquiring	2
Commitment	7	Leadership	11
Communication	13	Learn	12
Continuous	4	Management	5
Culture	7	Mutual Trust	4
Defer	4	Process	5
Empowerment	3	Responsiveness	3
Environment	8	Risk	4
Feedback	2	Safety	24
Hazard	2	Training	2

After comparison, the safety culture factors identified by BSEE are being strongly considered for this project. Their selection for consideration is based on several factors. First, they align very well with the research and reviews conducted for this project. Second, the BSEE safety factors are familiar to owners and operators in the US Gulf of Mexico. Third, we have the ability to build upon these factors to further refine and define the safety culture factors. Table 8 lists the candidate safety culture factors and provides example safety culture factor attributes and/or characteristics.

5. Conclusion and Future Work

Safety culture is a very important concept that reveals indicators of the safety culture perceptions of personnel within an organization. Often safety culture assessment procedures develop a series of safety culture factors to group the results of questionnaires and other forms of safety culture assessment. Grouping the data facilitates drawing conclusions and suggesting improvement opportunities. Our analysis showed that different organizations have employed different sets of factors to explain the results for safety culture. By counting the frequency of keywords, the factors, such as "learn", "leadership", "communication", etc., were identified and will be candidates receiving a higher priority when being considered for this project's toolbox's framework. The safety culture factors identified by BSEE safety culture policy are being strongly considered for this project.

Additional ongoing work includes the development of associated question sets, strategies for personnel interviews, methods for questionnaire distribution, collection, and analysis, dissemination of results, and guidance on implementation. The team looks forward to an opportunity to further share the results of this very important project and associated tools with the audience of this conference.

Table 8 Candidate Project Safety Culture Factors and associated characteristics or attributes

Safety Culture Factors	EXAMPLE - Factor Attributes / Characteristics
1. Leadership Commitment to Safety Values and Actions	– Allocate sufficient resources to safety – Consistent messaging (verbal and non-verbal) on the priority of safety in the organization – Fund important safety-related projects – Zero tolerance for willful violations – Actions taken under stop work authority are supported by management
2. Hazard Identification and Risk Management	– Frequently discuss process hazards – Hazard/risk assessments that are used for continuous improvement rather than just regulatory compliance – Training to recognize and report unsafe acts and conditions
3. Personal Accountability	– Personal commitment to safety from all personnel at all levels – Follow policies and procedures – Stop work authority is understood and accepted by all – Personnel accepts and fulfills individual responsibilities
4. Work Processes	– Establish a safety vision – Framework exists to define safety culture as a key element – Safety included in annual objectives – Operating and maintenance procedures are current and accurate – Maintenance backlog is within bounds set by internal standards – Recommendations/findings (e.g., MOC, audits) are quickly resolved
5. Continuous Improvement	– Continue vigilance even when performance indicators are positive – Prompt, thorough incident investigations – Procedure updates are quickly resolved and incorporated
6. Environment for Raising Concerns	– Near miss reporting system, with a periodic evaluation of trends – Root cause investigations go past the surface to better understand root causes (e.g., "failure to follow a procedure" requires further understanding of "why?") – "It's always been that way" mentality is challenged
7. Effective Safety and Environmental Communication	– Communicate internal and external incidents – Open 2-way channels of communication (vertical and horizontal) – Variety of means of communication used – Rapid communication and response to investigations and safety suggestions – Feedback always provided even if the suggestion is not adopted
8. Respectful Work Environment	– Subject matter experts are trained, developed, and recognized, including process safety expertise – Organization turns to the experts rather than rank – Investigations focus on solving the problem rather than on blame – Fair and just system exists so that honest errors can be reported without fear of reprisals
9. Inquiring Attitude	– Personnel are trained/retrained on their job responsibilities – Active behavior-based safety program

ACKNOWLEDGEMENTS

This research is partially sponsored by the National Academies of Sciences, Engineering, and Medicine grant #20001064 to American Bureau of Shipping, American Bureau of Shipping Group, Lamar University, and University of Houston.

Appendix I Acronyms of Associations or Agencies

Associations	Title
HRO	High-Reliability Organization
Reasons	James Reasons numerous books on Human Error
CCPS	Center for Chemical Process Safety
RBPS	Risk Based Process Safety
CCHS	Contra Costa County Health Services
Japan OGI	Japan Oil & Gas International
UK HSE	Health and Safety Executive (not open to public)
UK PSLG	Process Safety Leadership Group
ABS	American Bureau of Shipping
BSEE	Bureau of Safety and Environmental Enforcement
COS	Center for Offshore Safety
IOGP	International Association of Oil & Gas Producers
NOPSEMA	The National Offshore Petroleum Safety and Environmental Management Authority
PAME	Protection of the Arctic Marine Environment
CNEB	Canadian National Energy Board
INPO	Institute of Nuclear Power Operations
IAEA	International Atomic Energy Agency
NRC	US Nuclear Regulatory Commission
NTSB	National Transportation Safety Board
PHMSA	Pipeline and Hazardous Materials Safety Administration

REFERENCES

Barrett, A. L. (1981). The urban unease revisited: Perceptions of local safety and neighborhood satisfaction among metropolitan residents. Social Science Quarterly, 62(4), 611.

Broni-Bediako, E. &. (2010). Effects of drilling fluid exposure to oil and gas workers presented with major areas of exposure and exposure indicators. Research Journal of Applied Sciences, Engineering and Technology, 2(8), 710-719.

Center for Chemical Process Safety. (2007). *Guidelines for Risk Based Process Safety.* John Wiley & Sons.

Cheyne, A. C. (1998). Modelling safety climate in the prediction of levels of safety activity. Work & Stress, 12(3), 255-271.

Clarke, S. (1999). Perceptions of organizational safety: implications for the development of safety culture. Journal of Organizational Behavior: The International Journal of Industrial, Occupational and Organizational Psychology and Behavior, 20(2), 185-198.

Cooper, D. (1997). Evidence from safety culture that risk perception is culturally determined. International Journal of Project and Business Risk Management, 1(2): 185-202.

Cooper, M. D. (2000). Towards a model of safety culture. Safety science, 36(2), 111-136.

Cox, S. C. (1991). The structure of employee attitudes to safety: A European example. Work & stress, 5(2), 93-106.

Cox, S. J. (2000). J. T. Cheyne. "Assessing safety culture in offshore environments." Safety science 34, no. 1-3 (2000): 111-129.

Cox, S., & Flin, R. (1998). Safety culture: philosopher's stone or man of straw? *Work & stress, 12*(3), 189-201.

Coyle, I. R. (1995). Safety climate. Journal of Safety research, 26(4), 247-254.

Curcuruto, M. M. (2018). "Multilevel safety climate in the UK rail industry: A cross validation of the Zohar and Luria MSC scale." Safety science 110 (2018): 183-194.

Denison, D. R. (1996). What is the difference between organizational culture and organizational climate? A native's point of view on a decade of paradigm wars. Academy of management review, 21(3), 619-654.

Edwards, J. R. (2013). Returning to the roots of culture: A review and re-conceptualisation of safety culture. Safety science, 55, 70-80.

Fang, D. C. (2006). Safety climate in construction industry: A case study in Hong Kong. Journal of construction engineering and management, 132(6), 573-584.

Fogarty, G. J. (2010). Safety climate and the theory of planned behavior: towards the prediction of unsafe behavior. Accid Anal Prev, 42(5): 1455-1459.

Glendon, A. I. (2000). Perspectives on safety culture. Safety science, 34(1-3), 193-214.

Grabowski, M. Z. (2009). "Sailing on friday: Developing the link between safety culture and performance in safety-critical systems." IEEE Transactions on Systems, Man, and Cybernetics-Part A: Systems a.

Griffin, M. A. (2000). Perceptions of safety at work: a framework for linking safety climate to safety performance, knowledge, and motivation. J Occup Health Psychol, 5(3): 347-358.

Guldenmund, F. W. (2000). The nature of safety culture: a review of theory and research. Safety science, 34(1-3), 215-257.

Guldenmund, F. W. (2010). Understanding and exploring safety culture.

Hale, A. R. (2000). Culture's confusions. Safety science, 34(1-3), 1-14.

Håvold, J. I. (2017, September). Safety culture in a shipping company. Evidence from two surveys 13 years apart. In Prevention of Accidents at Work: Proceedings of the 9th International Conference on the Prevention of Accidents at Work (WOS 2017). CRC Press.

Health and Safety Commission. (1993). *Organizing for safety: Third report of the human factors study group of ACSNI. Sudbury, UK: HSE Books.*

Høivik, D. M. (2009). An explorative study of health, safety and environment culture in a Norwegian petroleum company. Safety science, 47(7), 992-1001.

Hon, C. K. (2014). "Relationships between safety climate and safety performance of building repair, maintenance, minor alteration, and addition (RMAA) works." Safety science 65 (2014): 10-19.

Hystad, S. W.-R. (2013). "Fatigue in seafarers working in the offshore oil and gas re-supply industry: effects of safety climate, psychosocial work environment and shift arrangement." International maritime healt.

INSAG. (1992). *INSAG-7 They Chernobyl Accident: Udpating of INSAG-1.* Retrieved from International Nuclear Safety Advisory Group: https://www-pub.iaea.org/MTCD/publications/PDF/Pub913e_web.pdf

Isla Díaz, R. &. (1997). Safety climate and attitude as evaluation measures of organizational safety. Accident Analysis & Prevention, 29(5): 643-650.

Kennedy, R. &. (1998). Development of a hazard and operability-based method for identifying safety management vulnerabilities in high risk systems. Safety Science, 30(3), 249-274.

Mearns, K. K. (2013). Development of a methodology for understanding and enhancing safety culture in air traffic management. Safety Science, 53(0): 123-133.

Mohamed, S. (2003). Scorecard approach to benchmarking organizational safety culture in construction. Journal of construction engineering and management, 129(1), 80-88.

Morrow, S. L. (2014). Exploring the relationship between safety culture and safety performance in US nuclear power operations. Safety Science, 69, 37-47.

National Academies of Sciences. (2016). *Engineering, and Medicine. (2016). Strengthening the Safety Culture of the Offshore Oil and Gas Industry. Transportation Research Board.*

National Academies of Sciences, E. a. (2016). Strengthening the Safety Culture of the Offshore Oil and Gas Industry. Transportation Research Board, 2016.

Neal, A. G. (2000). The impact of organizational climate on safety climate and individual behavior. Safety Science, 34(1–3): 99-109.

Nielsen, M. B.-R. (2013). "A brief safety climate inventory for petro-maritime organizations." Safety science 58 (2013): 81-88.

Niskanen, T. (1994). Safety climate in the road administration. Safety Science, 17(4), 237-255.

Oltedal, H. A. (2009). Local management and its impact on safety culture and safety.

Oltedal, H. A. (2011). Reporting practices in merchant shipping, and the identification of influencing factors. Safety Science, 49(2), 331-338.

Oltedal, H. A. (2011). Safety culture and safety management within the Norwegian-controlled shipping industry; State of art, interrelationships, and influencing factors.

Patankar, M. S. (2010). The safety culture perspective. In Human factors in aviation (pp. 95-122). Academic Press.

Pidgeon, N. (1997). The limits to safety? Culture, politics, learning and man–made disasters. Journal of contingencies and crisis management, 5(1), 1-14.

Pidgeon, N. F. (1991). Safety culture and risk management in organizations. Journal of cross-cultural psychology, 22(1), 129-140.

Reiman, T. a. (2012). "Leading indicators of system safety–monitoring and driving the organizational safety potential." Safety science 50, no. 10 (2012): 1993-2000.

Richter, A. &. (2004). Integration, differentiation and ambiguity in safety cultures. Safety science, 42(8), 703-722.

Schein, E. H. (1990). Organizational Culture: What it is and How to Change it. In Human resource management in international firms (pp. 56-82). Palgrave Macmillan, London.

Shirali, G. A. (2016). Quantitative assessment of resilience safety culture using principal components analysis and numerical taxonomy: A case study in a petrochemical plant. Journal of Loss Prevention in the Process Industries, 40, 277-284.

Skogdalen, J. E. (2010). Developing safety indicators for preventing offshore oil and gas deepwater drilling blowouts. Safety science, 49(8-9), 1187-1199.

Stemn, E. B. (2018). Examining the relationship between safety culture maturity and safety performance of the mining industry. Safety Science, 113, 345–355. doi:10.1016/j.ssci.2018.12.008.

Tang, D. K. (2017). Factors affecting safety of processes in the Malaysian oil and gas industry. Safety Science, 92, 44–52. doi:10.1016/j.ssci.2016.09.017.

Toolkit, H. a. (2004). *Hearts and Minds Toolkit: Learning from incidents questionnaire.* HSE 184.

Turner, B. A. (1989, November). Safety culture: its importance in future risk management. In Position paper for the second World Bank workshop on safety control and risk management, Karlstad, Sweden (pp. 6-9).

Wiegmann, D. A. (2002). A synthesis of safety culture and safety climate research. Savoy: Aviation Research Lab, Institute of Aviation.

Zahoor, H. A. (2017). "Modeling the relationship between safety climate and safety performance in a developing construction industry: a cross-cultural validation study." International journal of environ.

Zahoor, H. C. (2017). Determinants of safety climate for building projects: SEM-based cross-validation study. Journal of construction engineering and management, 143(6), 05017005.

Zohar, D. (1980). Safety climate in industrial organizations: theoretical and applied implications. Journal of applied psychology, 65(1), 96.

SNAME
THE INTERNATIONAL COMMUNITY
FOR MARITIME AND OCEAN PROFESSIONALS

PREDICTING FUEL CONSUMPTIONS AND EXHAUST GAS EMISSIONS FOR LNG CARRIERS VIA MACHINE LEARNING WITH HYPERPARAMETER OPTIMIZATION

CHENXI JI
THE GAS & FUELS RESEARCH CENTER, TEXAS A&M UNIVERSITY

Proceedings of the 26th Offshore Symposium, April 6 & 7 2021, Houston, Texas
Texas Section of the Society of Naval Architects and Marine Engineers

ABSTRACT

The prediction of marine fuel consumption and ship exhaust gas emissions are indispensable to evaluating ship sustainable performance under current shipping fuel standards. Big data with evolved machine learning techniques have been proved to be an effective way to contain uncertainties for ship activities. This work collects the latest global LNG carrier fleet with 435 data points and attempts to predict the marine fuel consumptions and ship-resulted global warming potential (GWP) gas emissions, including CO_2, CH_4, N_2O, and black carbon aerosols. Gaussian process regression and ensemble machine learning approaches, to achieve this goal, are employed to infer the relationship between predictors (i.e., dimensional parameters, machinery parameters, and tonnage) and response variables (fuel consumptions and GWP exhaust gas emissions), providing exceptional insight into ship sustainable solutions. To improve the prediction accuracy, the hyperparameter optimization analysis via random search and Bayesian optimization is adopted to find the optimal machine learning model. The appealing results are in line with the validation data, illustrating high effectiveness and robustness of the proposed machine learning models. The procedure established in this study presents a novel approach for accelerating the research and development of sustainable shipping fuels under normal ship activities.

Keywords: Shipping sustainability; Machine learning; Bayesian optimization; Fuel consumption; GWP gas emissions

INTRODUCTION

Many public concerns have been brought to the increasingly intense greenhouse effects. As a specialized agency of the United Nations, the International Maritime Organization (IMO) has set its long-term decarbonization strategy to achieve its mission of "safe, secure, clean and sustainable shipping". In April 2018, an Initial Greenhouse Gas Strategy was proposed to contain the carbon intensity by 40% by 2030, and 70% by 2050 (MEPC 72, 2018). IMO has completed four shipping emission and fuel consumption studies for the recent two decades, pointing to the strong relationship between fuel oil consumption and ship exhaust gas emissions. Also, ships of 5,000 gross tonnage and above were enforced to provide the fuel oil consumption data to the ship's flag state and the methodology and process used to collect data needs to be clarified as per the Ship Energy Efficiency Management Plan (SEEMP) (International Maritime Organisation, 2016). Later in 2020, the Annex VI of the International Convention for the Prevention of Pollution from Ship (MARPOL), "2020 Sulfur Cap", was enforced to globally reduce the shipping fuel sulfur content on board from 3.5% mass to 0.5% mass outside the emission control areas. In response to the current marine environmental challenges, four shipping fuel solutions, including heavy fuel oil (HFO) with scrubber, marine gas oil (MGO), very low sulfur fuel oil (VLSFO), and Liquefied Natural Gas (LNG) are widely accepted in the shipping community after considering the infrastructure availability and technology maturity. LNG, deemed as the cleanest fossil fuel, is a promising fuel option to comply with the 2020 sulfur cap and the 2050 decarbonization strategy.

In general, there are three ways to estimate the fuel consumption and ship-induced emissions. The direct data collecting approach covers the bunker delivery note (BDN), tank level indicators (TLI), and flow metering (Hunsucker et al., 2018), and the uncertainties usually lay down the operation performance. The top-down approach is a method to calculate the total emissions without considering the vessel characteristics and the estimated values are assigned to the different ships in a specific fleet or area (DNV GL, 2019; EMEP/EEA, 2019; ENTEC, 2010); It is an accessible way to evaluate regional emission inventories, but the ship data tend to have a large number of uncertainties because of ocean-going vessels traveling among continents, and the fuel oil characteristics and prices cannot be consistent from port to port as well. A more realistic approach is the bottom-up approach which relies on calculating the emission by a ship in a specific position, and the total emissions are estimated by aggregating these estimates over time and the fleet (EPA, 2000; Eyring et al., 2005; Smith et al., 2014). Owing to the easily accessed ship data provided by shipboard automatic information system (AIS), various studies (Eyring et al., 2005; Ji and El-Halwagi, 2020; Kwon et al., 2019; Moreno-Gutiérrez et al., 2019; Rakke, 2016; Smith et al., 2014; Starcrest Consulting Group, 2014; Wan et al., 2019) have employed the bottom-up method and many mathematical models have been proposed with promising robustness. Besides the shipping fuel consumption, this work will focus on the emissions of global warming potential (GWP) gases, covering CO2, CH_4, N_2O, and black carbon aerosols.

Most studies sought for optimizing the mechanism models provided by EPA (EPA, 2000), so the prediction quality of outputs largely relies on the accuracy of ship engine operation time, engine actual delivered power, and the specific fuel oil consumption (SFOC) (Ji and El-Halwagi, 2020). Machine learning (ML) models, which have been well investigated to predict liquid inherent safety factors (Jiao et al., 2020; Yuan et al., 2019) and to find optimal fuel safety performance (Ji et al., 2021a, 2021b; Yuan et al., 2020), are good match to expand this ship sustainability database. Some studies applied basic supervised ML models for ship exhaust gas emissions: a multi-linear regression method was proposed (Uyanik et al., 2019) to estimate the main engine fuel rate, but the predicted outputs displayed high collinearity between variables of the main engine revolutions per minute and shaft power; Yuan and Nian (Yuan and Nian, 2018) utilized seven independent variables, including speed, mean draft, trim, wind speed, wind direction, wave height, and wave direction, to construct their Gaussian process model with acceptable outputs. However, none of them has adopted the hyperparameter optimization techniques for the proposed ML models.

This manuscript integrated the bottom-up approach with optimization induced machine learning models for the first time to predict the ship fuel consumption per hour and GWP (also called CO_2 equivalent) gas emissions for global LNG fleet at cruising mode. 4 Gaussian process regression (GPR) and 2 ensemble regression models are processed without optimization method, then two hyperparameter optimization algorithms are incorporated with GPR and ensemble models respectively.

METHODOLOGY

Fuel consumption and GWP gas emission mechanism models

The paper works on the cruising mode emissions for marine air pollutants when the selected LNG carriers (LNGC) are navigating in the Non-Emission Control Areas (Non-ECAs), since it is a common practice to adopt auto-pilot mode when the ship is navigating in the stable external environment at the deep sea. Three emissions sources, including main engine, auxiliary engine and auxiliary boiler, were considered in this work. The emissions calculation for one ship fleet with each specific fuel option and exhaust gas pollutant can be determined by Equation 1:

$$\boldsymbol{Em}_{ph_{(i,j,k)}} = \sum\left[(\boldsymbol{P}_{ME_{MCR_{(i,j,k)}}} \cdot \boldsymbol{LF}_{ME_{(i,j,k)}} \cdot \boldsymbol{EF}_{ME_{(i,j,k)}} + \boldsymbol{P}_{AE_{MCR_{(i,j,k)}}} \cdot \boldsymbol{LF}_{AE_{(i,j,k)}} \cdot \boldsymbol{EF}_{AE_{(i,j,k)}} + \boldsymbol{P}_{AB_{ep_{(i,j,k)}}} \cdot \boldsymbol{EF}_{AB_{(i,j,k)}}) \cdot \mathbf{1\ hr}\right] \quad (1)$$

where i: Ship class, j: Fuel option, k: Air pollutant, Em_{ph}: Emissions for ship class i, fuel option j and air pollutant k per hour, $P_{ME_{MCR}}$: main engine maximum continuous rating power, LF_{ME}: main engine load factor, EF_{ME}: main engine energy-based emission factor, $P_{AE_{MCR}}$: auxiliary engine maximum continuous rating power, LF_{AE}: auxiliary engine load factor, EF_{AE}: auxiliary engine energy-based emission factor, $P_{AB_{ep}}$: auxiliary boiler effective power demand, EF_{AB}: auxiliary boiler energy-based emission factor.

Referred from several previous studies (ENTEC, 2010; Ji and El-Halwagi, 2020; Moreno-Gutiérrez et al., 2019; Olmer et al., 2017), this work updated the main engine load factor with an adjusted ship speed, hull roughness, weather adjusted factors and draught coefficient; the auxiliary engine load factor was determined by the ratio of main engine and auxiliary engine (Moreno-Gutiérrez et al., 2015). Therefore, the CO2 emission mechanism model is converted as follows:

$$\boldsymbol{Em}_{ph_{(LNGC,HFO,CO2)}} = \sum\left[(\boldsymbol{P}_{ME_{MCR_{(LNGC,HFO,CO2)}}} \cdot \left(\frac{\boldsymbol{Vact}}{\boldsymbol{V_{MCR}}}\right)^3 \cdot \boldsymbol{HFF}_{LNGC} \cdot \boldsymbol{WAC} \cdot \boldsymbol{DAF}_{LNGC} \cdot \boldsymbol{EF}_{ME_{(LNGC,HFO,CO2)}} + \boldsymbol{P}_{AE_{MCR_{(LNGC,HFO,CO2}}} \cdot \boldsymbol{LF}_{AE_{(LNGC,HFO,CO2)}} \cdot \boldsymbol{EF}_{AE_{(LNGC,HFO,CO2)}} + \boldsymbol{P}_{AB_{ep_{(LNGC,HFO,CO2)}}} \cdot \boldsymbol{EF}_{AB_{(LNGC,HFO,CO2)}}) \cdot \mathbf{1\ hr}\right] \quad (2)$$

where Vact is the actual ship speed, which equals to the speed over ground multiplying the speed adjusted factor; VMCR is the speed when ME is in maximum continuous rating mode, which is assumed as 1.064 times the normal continuous rating speed because the cruise speeds proposed by Lloyd's data are 94% of the maximum speed; HFF is the hull fouling factor, which is determined by the ship age, length between perpendiculars and the coefficient of marine biofouling; WAC, weather adjustment coefficient, and DAF, draught adjustment factor. Referred from two previous studies (Ji and El-Halwagi, 2020; Olmer et al., 2017), the emission factor for the four main engine exhaust gases are listed in Table 1. Also, the engine types of LNG carriers and the latest fuel option, VLSFO, have been integrated to the emission factor data.

Table 1: Emission factors (g/kwh) of currently available fuel solutions (referred from ENTEC, 2010; Ji and El-Halwagi, 2020; Moreno-Gutiérrez et al., 2019; Olmer et al., 2017)

	EF CO_2 (ME/AE/AB)	EF CH_4 (ME/AE/AB)	EF N_2O (ME/AE/AB)	EF BCA (ME, g/kg fuel)
HFO 2.5% S(SSD)	607/707/950	0.01/0.01/0.002	0.03/0.04/0.05	0.14
HFO 2.5% S(MSD/HSD)	670/707/950	0.01/0.01/0.002	0.03/0.04/0.05	0.14
HFO 2.5% S(GT/ST)	950/707/950	0.01/0.01/0.002	0.05/0.04/0.05	0.005 (GT)/ 0.08 (ST)
VLSFO 0.1~0.5% S	533/557/962	0.0052/0.0055/0.002	0.027/0.028/0.04	0.19
LNG-Diesel	366/366/366	0.94/0.94/0.94	0.011/0.011/0.011	0.002

Many leading factors, such as engine types, fuel options, may contribute to the fuel oil consumptions. CO_2 intensity of HFO was introduced to calculate fuel oil consumption from CO_2 emissions per hour, shown as $FC_{ph_{LNGC,HFO}}$ in equation 3:

$$\boldsymbol{FC_{ph_{LNGC,HFO}}} = \sum_{HFO}\left(\frac{Em_{CO_{2ME\ LNGC,HFO}} + Em_{CO_{2AE\ LNGC,HFO}} + Em_{CO_{2AB\ LNGC,HFO}}}{CO_2\ Intensity}\right) \quad (3)$$

where $Em_{CO_{2ME\ LNGC,HFO}}$, $Em_{CO_{2AE\ LNGC,HFO}}$, $Em_{CO_{2AB\ LNGC,HFO}}$ denote CO2 emissions for main engine, auxiliary engine and auxiliary boiler, respectively. The Global Warming Potential gases, firstly proposed by Intergovernmental Panel on Climate Change (IPCC), usually denote CO_2, CH_4, N_2O, Chlorofluorocarbons (CFCs), hydrofluorocarbons (HFCs), hydrochlorofluorocarbons (HCFCs), perfluorocarbons (PFCs), and sulfur hexafluoride (SF6) (Sabine, 2014). Several studies (Bond et al., 2013; Comer et al., 2017; Ji and El-Halwagi, 2020) pointed out that the BCA from ship exhaust gases have contributed a lot for GWP, and based on the calculated outputs of Comer's research team, the BCA accounted for more than 20% for 20 year-GWP and 7% for GWP_{100}. Thus, this paper adopted the IPCC fifth report GWP_{100} value for CO_2, CH_4 and N_2O, and the BCA GWP_{100} was referred from Bond's study (Bond et al., 2013). Equation 4 shows the way to calculate the emission of CO_2 equivalent by considering the emissions of BCA:

$$\sum \boldsymbol{Em_{CO_{2e(LNGC,HFO)}}} = \sum \boldsymbol{Em_{CO_{2(LNGC,HFO)}}} + \boldsymbol{GWPC_{CH_4}} \cdot \sum \boldsymbol{Em_{CH_{4(LNGC,HFO)}}} + \boldsymbol{GWPC_{N_2O}} \cdot \sum \boldsymbol{Em_{N_2O_{(LNGC,HFO)}}} + \boldsymbol{GWPC_{BCA}} \cdot \sum \boldsymbol{Em_{BCA_{(LNGC,HFO)}}} \quad (4)$$

where $Em_{CO_{2e(LNGC,HFO)}}$: GWP equivalent CO_2 emission for LNG carrier with HFO, Em_x: Emission of CO_2, CH_4, N_2O and Black Carbon Aerosol, $GWPC_x$: Global Warming Potential Coefficient of CO_2, CH_4, N_2O and Black Carbon Aerosol.

Data set and response variables

The sustainable LNGC database (SLD) used in this study has consisted of all the static and dynamic global LNG tanker fleet collected from the Lloyd's Maritime Intelligence Unit (LMIU). Then all the LNGC data were checked for missing values and the engine data were recalculated if dual engines were on board. After cleaning and pre-processing the data in the significant variables, the final database contains the ship dimensional, machinery and emission data from 435 LNG tankers.

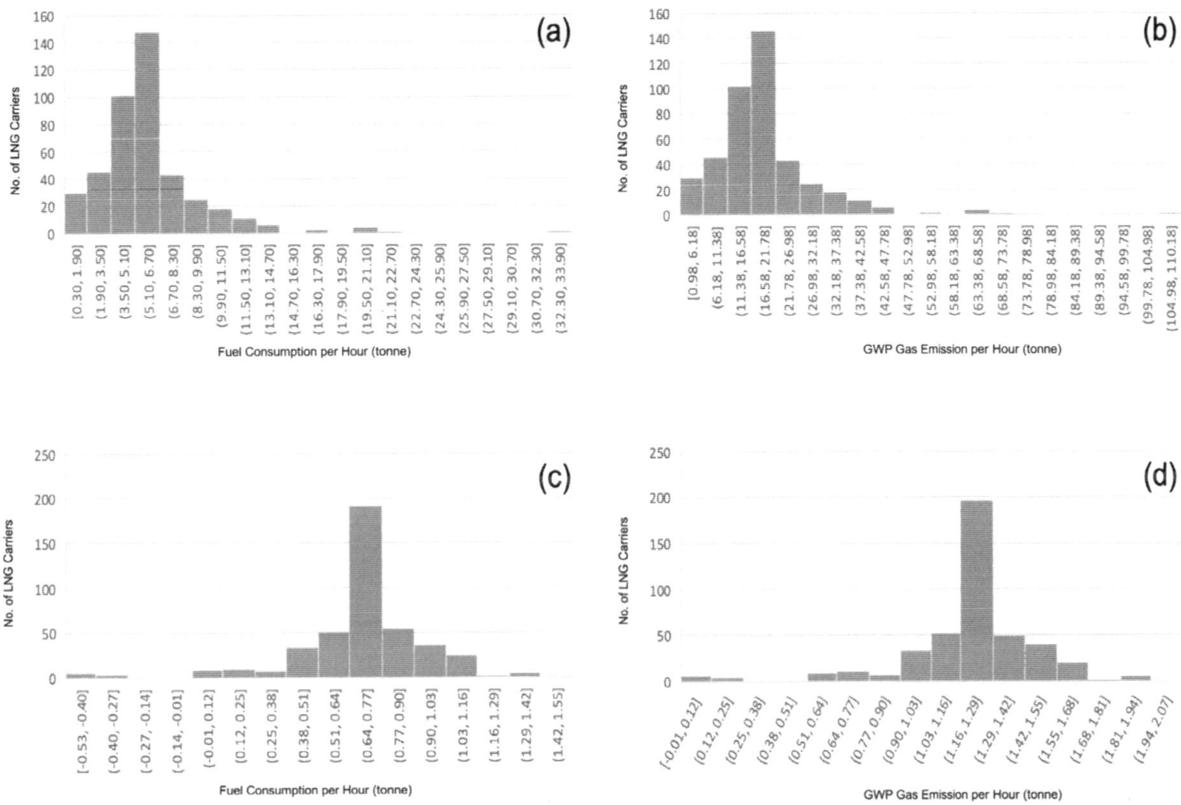

Figure 1: a. Fuel consumption data distribution for SLD; b. GWP gas emission data distribution; c. Fuel consumption data distribution after logarithmic transformation; d. GWP gas emission data distribution after logarithmic transformation.

As indicated by Figure 1a and 1b, the distribution of Fuel Consumption (FC) and GWP gas emission display a right-skewed trend, especially for GWP gas emission, its distribution highly centralized close to zero with many noises on the right side. Then logarithmic transformations were executed for the two response variables to obtain an acceptable distribution of the dependent parameters, see Figure 1c and 1d. Then the SLD was randomly split into two sets, 350 data points accounted for training set while 85 data points were for testing data to avoid overfitting.

Feature selection and predictor correlations

30 independent variables, covering both static and dynamic parameters, are set as input to find the most statistically significant predictors. Here, the backward stepwise approach is adopted to process feature selection, and the AIC and BIC values are utilized as the criteria to determine the number of predictors. Two seven-feature models (including intercept) were selected as the minimum AIC/BIC values with the least number of predictors. The two models both point out the significant predictors as power, ship ages, molded breadth, liquid tank capacity, service speed and net tonnage. As shown in Figure 2, the FC model leading factors follow the rank of power > ship ages > molded breadth > liquid tank capacity > service speed > net tonnage, while the dominated predictors for GWP model follow power > ship ages > liquid tank capacity > molded breadth > service speed > net tonnage.

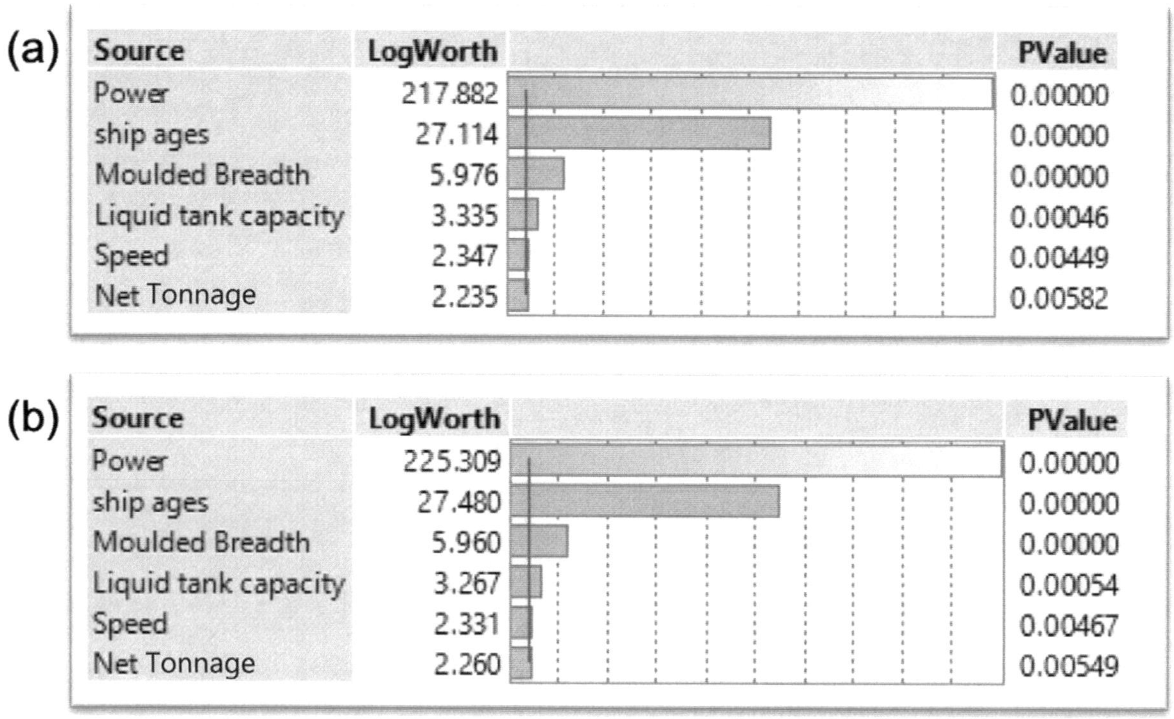

Figure 2: a. Statistical significance of identified predictors for FC; b. Statistical significance of identified predictors for GWP

Then the potential flaws for predictor assumptions such as non-constant error variance, outliers, abnormal leverage points and collinearity between independent variables were checked to ensure they have statistical significance. Also, the covariance and correlation between any two predictors $(x_1, \dots, x_n)$ are calculated following the below equations:

$$Cov(x_j, x_k) = \frac{1}{n-1}\sum_{i=1}^{n}(x_{j_i} - \overline{x_j})(x_{k_i} - \overline{x_k}) \quad (5)$$

$$r_{x_j x_k} = r_{x_k x_j} = \frac{Cov(x_j, x_k)}{s_{x_j} s_{x_k}} \quad (6)$$

where $Cov(x_j, x_k)$ and $r_{x_j x_k}$ are the covariance and correlation between x_j and x_k, s_{x_j} and s_{x_k} denote the standard deviations for x_j and x_k, respectively. Figure 3 shows the scatterplot matrix among the six predictors. Overall, all the predictors follow approximately normal distributions. There might be a collinearity issue between Net Tonnage and Molded Breath. In general, there are three ways to deal with highly collinear items: remove the highly correlated predictor directly, perform a principal component analysis (PCA) to reduce predictor numbers, add extra linearly combined pairwise independent variables. Compared with the three transformed models, the work kept the six-predictor model since it displayed the best overall bias-variance tradeoff performance and the high collinearity variable "Net Tonnage" has least significance for both FC and GWP predictions, shown in Figure 2.

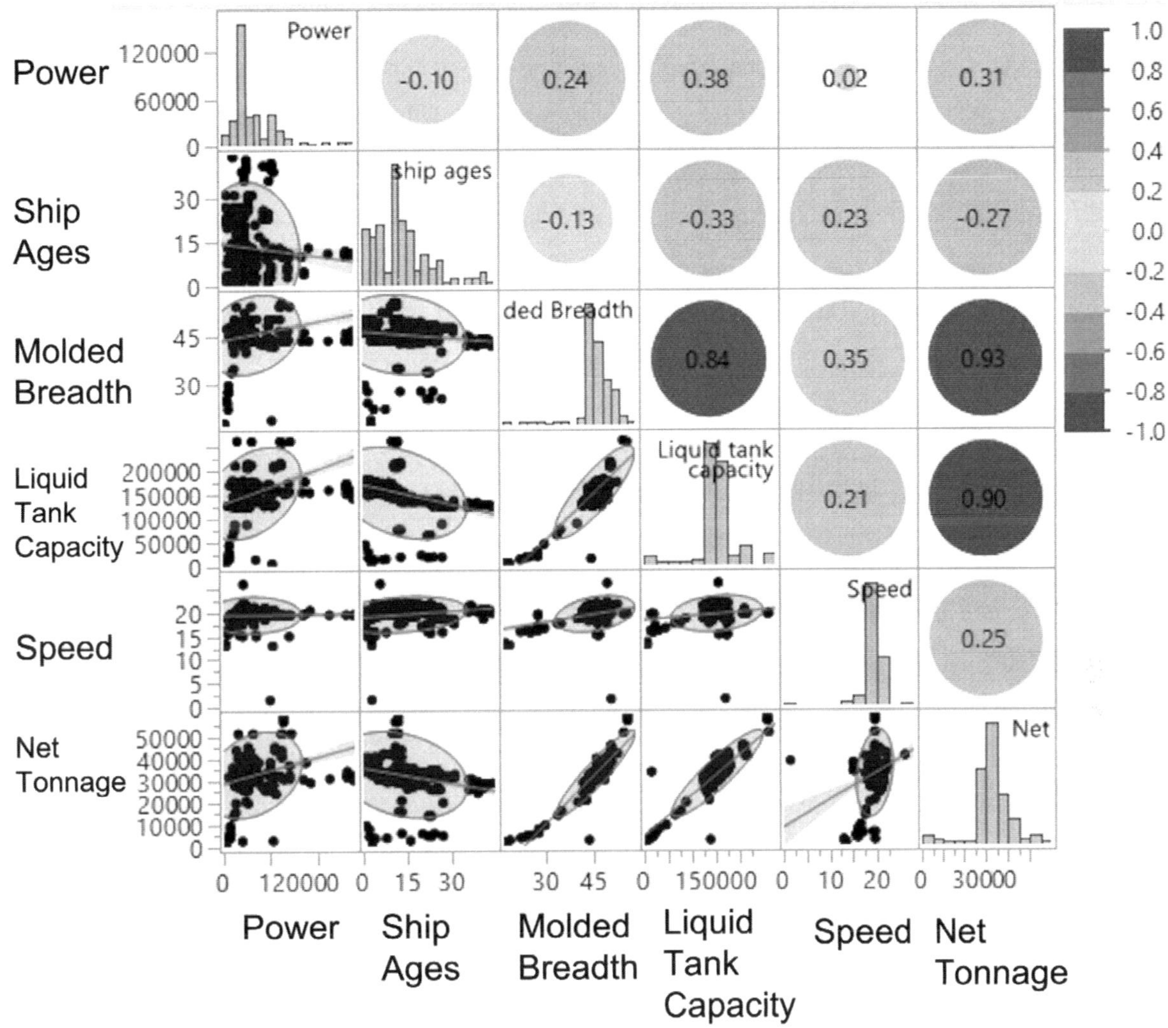

Figure 3: Scatter plot matrix with pairwise correlation heat map

Machine learning algorithms and hyperparameter optimization

The basic idea of GPR is to employ a linear model with normalized noise to get the predictive distribution by coupling with the likelihood function and posterior probability function (Rasmussen and Williams, 2018), and the GPR function and the predictive distribution function are shown in the equations below.

$$y_i = X_i^T w + \varepsilon, \varepsilon = N(0, \sigma_n^2 I) \; i = 1, \dots, n. \tag{7}$$

$$p(\tilde{y}|X, y, \tilde{X}) = N(\hat{\mu}, \hat{\Sigma}) \tag{8}$$

$$\hat{\mu} = K(\tilde{X}, X)^T (K(X, X) + \sigma_n^2 I)^{-1} (y - \mu(\tilde{X})) \tag{9}$$

$$\hat{\Sigma} = K(\tilde{X}, \tilde{X}) - K(\tilde{X}, X)^T (K(X, X) + \sigma_n^2 I)^{-1} K(\tilde{X}, \tilde{X}) \tag{10}$$

Following the above functions, $y_i = (y_1, y_2, \dots, y_n)^T$ denotes the response variables, X_i^T are the independent variables, ε is the Gaussian noise, K is the $n \times n$ covariance matrix, μ is the mean vector matrix, $\tilde{X}$ it the test locations, $\mu(\tilde{X}) = [\mu(x_{n+1}), \dots, \mu(x_{n+m})]^T$, $K(X, X) = K$, and $K(\tilde{X}, \tilde{X})$ is an $m \times m$ matrix. The only difference among the rational quadratic GPR, squared exponential GPR, exponential GPR and Matern 5/2 GPR is the kernel functions.

Two ensemble learning paradigms were employed in this study: Bagging, often called Bootstrap Aggregating, and Boosted decision trees. Bagged trees are processing random sampling from the original sample. In each round, n samples (Bootstrap method) with replacement from the N original samples are taken as the training set, and k rounds of extraction are performed to obtain k training sets (the training sets are independent of each other). Each time a training set is used to train a model, k training sets are used to obtain a total of k base models. The calculated average values of k base models are presented to solve regression problems. The bagged algorithm mainly focuses on decreasing variance other than bias, so it is fit for a high variance and low bias model (Fortmann-Roe, 2012).

The core idea of boosting trees is to assemble multiple weak learners into a strong one. It is a sequential ensemble approach. The base model generated in each iteration mainly improves the poor performance of the previously generated base model. The low variance and high bias models are preferred when considering applying a boosting tree paradigm.

There is no denying that the developed ML models can be optimized by tuning them by adjusting the internal parameters of the model. Hyperparameter optimization is a technique to automatically select the desired values to minimize the mean squared error (Feurer and Hutter, 2019). The optimizable GPR model integrates the eight kernel functions, using both isotropic and non-isotropic kernel functions to pre-process GPR ML models, while the kernel scale is limited with the range between 0.01 and the maximum predictor data, which is defined as the difference between the largest and least predictor values. The optimizable ensemble searches the minimum leaf size, number of learners, number of predictors to sample, and learning rate, which is among real values log-scaled in the range of [0.001, 1].

Model evaluation

R squared value is a common criterion to determine how much percentage of the training data can be represented by the proposed model, but it has a shortage since its value is keeping increasing as the number of predictors grows. This manuscript, thus, employs adjusted R squared value, shown in Equation 9, to avoid redundant insignificant predictors involving in the optimized FC and GWP models:

$$\boldsymbol{R^2_{adj.}} = \mathbf{1} - \left(\frac{n-1}{n-p-1}\right)\left[\mathbf{1} - \frac{\sum_{i=1}^{n}(\widehat{y_i}-\bar{y})^2}{\sum_{i=1}^{n}(y_i-\bar{y})^2}\right] \quad (11)$$

where n is the number of data points in the sample, p is the number of predictors, $\widehat{y_i}$ is the predicted value, y_i is the observed value and $\bar{y}$ is the average of the observed values. On the other side, the root means squared error (RMSE), mean squared error (MSE) and mean absolute error (MAE) are chosen to be the performance values for statistical errors. As k-fold cross validation (k=10) is adopted in this work, the modified formulae are listed below:

$$\boldsymbol{RMSE}_{k-Fold} = \frac{1}{k}\left(\sum_{n=1}^{k}\sqrt{\frac{\sum_{i=1}^{n}(\hat{y}_{ki}-y_{ki})^2}{n}}\right) \quad (12)$$

$$\boldsymbol{MSE}_{k-Fold} = \frac{1}{k}\left(\sum_{n=1}^{k}\frac{\sum_{i=1}^{n}(y_{ki}-\hat{y}_{ki})^2}{n}\right) \quad (13)$$

$$\boldsymbol{MAE}_{k-Fold} = \frac{1}{k}\left(\sum_{n=1}^{k}\frac{\sum_{i=1}^{n}|\hat{y}_{ki}-y_{ki}|}{n}\right) \quad (14)$$

where y_{ki} is the observed value in the kth iteration, $\hat{y}_{ki}$ is the corresponding predicted value of k-fold cross validation. Thus, the best prediction model is selected by balancing its predictability and interpretability, the least RMSE value and the largest adjusted R square value, as well as the shortest training time for the hyperparameter optimization approaches.

RESULTS AND DISCUSSION

Table 2: Prediction performance parameters of GPR and ensemble models for FC and GWP

No.	Machine Learning Model	RMSE	Adjusted R^2	MSE	MAE
Fuel Consumption					
1	Rational Quadratic GPR	0.056	0.94	0.003	0.02
2	Squared Exponential GPR	0.068	0.92	0.005	0.03
3	Matern 5/2 GPR	0.16	0.68	0.03	0.04
4	Exponential GPR	0.14	0.76	0.02	0.04
5	Ensemble Boosted Trees	0.071	0.94	0.005	0.05
6	Ensemble Bagged Trees	0.17	0.66	0.03	0.10
GWP Gas Emission					
1	Rational Quadratic GPR	0.064	0.93	0.004	0.02
2	Squared Exponential GPR	0.096	0.84	0.009	0.03
3	Matern 5/2 GPR	0.17	0.62	0.03	0.05
4	Exponential GPR	0.14	0.76	0.02	0.04
5	Ensemble Boosted Trees	0.10	0.87	0.01	0.07
6	Ensemble Bagged Trees	0.16	0.68	0.03	0.09

Considering the acceptable criterion of $R^2 > 0.6$ and RMSE < 0.5 (Witten et al., 2011), all the proposed ML models in Table 2 for the two response variables are tested to be satisfied ones. In particular, the GPR model with best fuel consumption prediction performance is Rational Quadratic GPR, while that for ensemble model is Boosted Trees. For GWP gas emission prediction, the optimal ML models are Rational Quadratic GPR and Boosted Trees as well. However, the hyperparameter optimization were conducted in this work to seek for further accurate ML models with better prediction performance. Two approaches, random search and Bayesian optimization, were employed for all the proposed GPR and ensemble models. From the outputs of Table 2, the RMSE values and R square values were evidently demonstrated to be improved as a result of hyperparameter optimization.

Among all the four optimized ML models, the random search ensemble one presents the lowest RMSE and highest r square value for fuel consumption prediction, and there is a 34% decrease for RMSE compared to the optimal ML model without optimization techniques. Similarly, the optimizable GWP gas emission model is selected as a Bayesian optimization ensemble with the RMSE value of 0.046 and R square value of 0.97. Besides the prediction accuracy performance parameters, this work also integrated two prediction efficiency parameters, prediction speed and training time. As shown in Table 3, the prediction speed of Bayesian optimization GPR tops other models for both fuel consumption and GWP gas emissions; The two optimizable ensemble approaches save as much as 1100 seconds for training models as compared to the two optimized GPR methods. Interestingly, the most accurate models for the two response variables also display the best time-saving performance.

Table 3: Optimizable prediction performance for fuel consumption and GWP gas emissions

	Hyperparameter Optimization ML Model	RMSE	Adjusted R^2	MSE	MAE	Prediction Speed (obs/sec)	Training Time (sec)
Fuel Consumption							
Optimizable GPR	Random search GPR	0.046	0.96	0.002	0.02	3500	1094.2
	Bayesian optimization GPR	0.045	0.96	0.002	0.02	8000	1436.3
Optimizable Ensemble	Random search Ensemble	0.037	0.98	0.001	0.02	520	281.2
	Bayesian optimization Ensemble	0.050	0.97	0.002	0.03	710	323.1
GWP Gas Emission							
Optimizable GPR	Random search GPR	0.049	0.96	0.002	0.02	9100	561.8
	Bayesian optimization GPR	0.063	0.93	0.004	0.02	11000	394.6
Optimizable Ensemble	Random search Ensemble	0.063	0.95	0.004	0.03	500	303.5
	Bayesian optimization Ensemble	0.046	0.97	0.002	0.03	1800	290.1

Optimizable fuel consumption model performance

A total number of 6 machine learning models, including 4 GPR models and 2 ensemble trees have been developed for the prediction of fuel consumption as the first stage. The Rational Quadratic GPR model with the lowest RMSE value (0.056), MAE (0.02), MSE (0.003) and the largest R square value (0.94), is found to be the best one among all GPR models; Also, it outperforms the optimal ensemble ML model, boosted tree, of which the RMSE and R square value are 0.071 and 0.94, respectively. Then, two hyperparameter optimization techniques, random search and Bayesian optimization, were executed for GPR models and ensemble ones separately. As illustrated in Table 3, the favorable RMSE values for the four optimizable ML models follow the rank of Random search Ensemble > Bayesian optimization GPR > Random search GPR > Bayesian optimization Ensemble. Furthermore, the Random search

Ensemble model presents the largest R square value (0.98) and the shortest training time (281.2s), so it is set to be the preliminary optimal model. Figure 4a shows that there is no evidence of heteroscedasticity of the preliminary model, demonstrating the model was in line with the requirement of constant error variance; The assumption of statistical significance was met since none of residual plots was located outside the acceptable range (-0.2, 0.2) (Ji et al., 2021b; Pinho et al., 2015). Meanwhile, the overall predicted vs experimental response plot, shown in Figure 4b, denotes that the data points were evenly distributed with high-quality prediction accuracy.

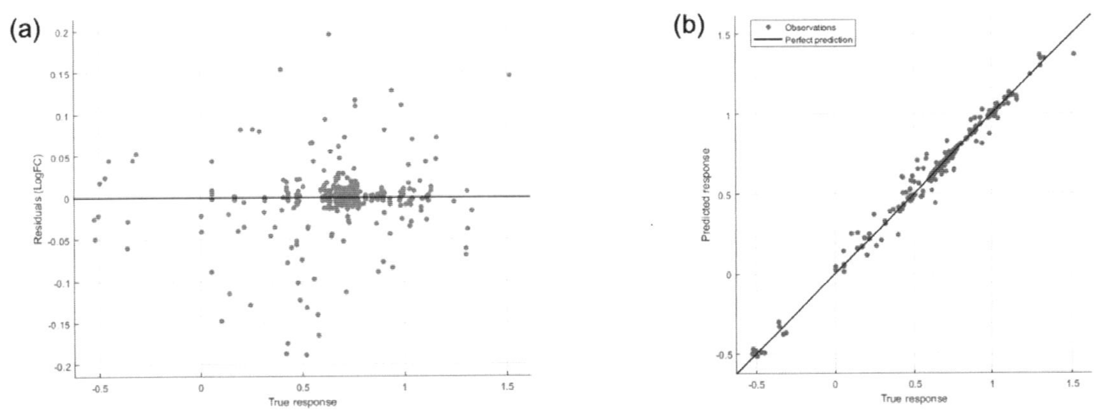

Figure 4: a. Residual plot for optimizable ensemble FC; b. Predicted vs experimental response plot for optimizable ensemble FC

The random search ensemble based hyperparameter optimization approach was limited to 30 iterations, and the range of learner numbers is between 10 to 500 with learning rate of 0.001~1; while the minimum leaf size is set in the range of (1, 219). The optimization results listed in Figure 5 present that the optimal ensemble method is LSBoost tree with 416 learners, minimum leaf size of 3, learning rate of 0.036 and 4 predictors to sample. More importantly, the best point hyperparameters and minimum error hyperparameters occurred simultaneously when iteration 5 was taken.

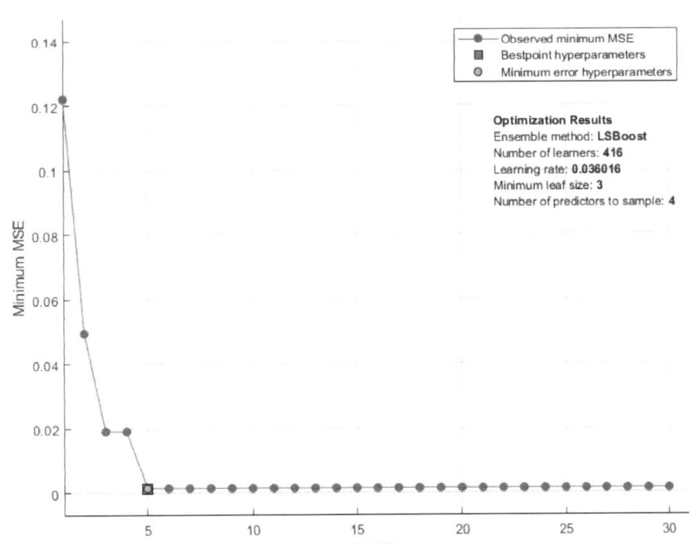

Figure 5: Random search hyperparameter optimization performance for ensemble model of FC

Optimizable GWP gas emission model performance

Firstly, the best prediction ML models without optimization for GPR and ensemble algorithms are determined as rational quadratic GPR and LSBoost, respectively. Overall, the rational quadratic GPR is the optimal ML model without hyperparameter optimization with the lowest RMSE value (0.064), MAE (0.02), MSE (0.004) and the largest R square value (0.93). The hyperparameter optimization was processed with the outputs illustrated in Table 3, RMSE values, from high to low, follow the rank of Random search Ensemble = Bayesian optimization GPR > Random search GPR > Bayesian optimization Ensemble. Also, the Bayesian optimization Ensemble model presents the largest R square value (0.97) and the most favorable training timesaving (290.1s).

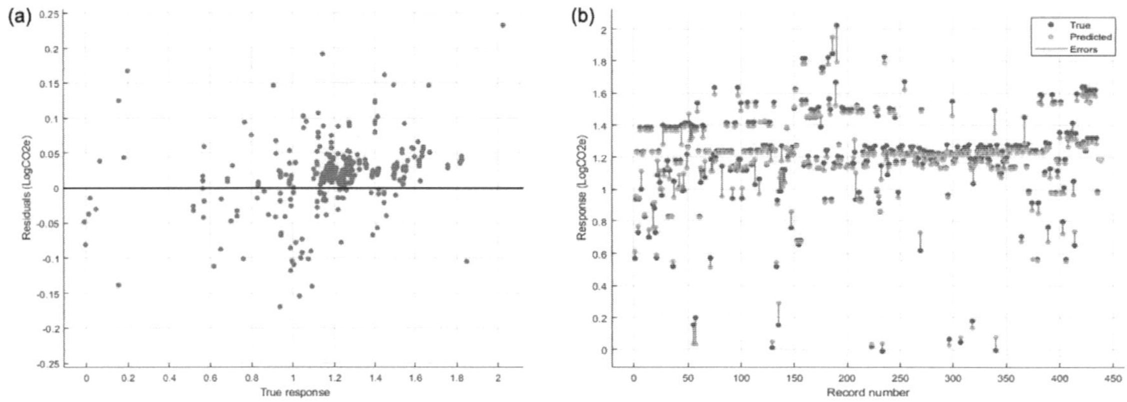

Figure 6: a. Residual plot for optimizable ensemble GWP; b. Predicted vs experimental response plot for optimizable ensemble GWP

Next, the residual plots were checked to confirm the requirement of constant error variance and the assumption of statistical significance was met since there is only one outlier, displayed in Figure 6a, laying outside the range of (-0.2, 0.2), and it is neglectable since the sample consists of 435 data points. Presented in Figure 6b, the response plots for the optimized ML model for GWP gas emissions with the blue dots serving as the experimental data and yellow dots for predicted values show the prediction errors for each data point in the database. Moreover, 30 iterations were conducted for the Bayesian hyperparameter optimization. The hyperparameter search ranges are as follows: Number of learners (10, 500), learning rate (0.001, 1), minimum learning rate (1, 219), and the number of predictors to sample (1, 6). The final optimization model was generated when the minimum leaf size, number of learners, learning rate, and number of predictors equal to 1, 101, 0.04, and 4, respectively. As shown in Figure 7, the best point hyperparameter was finalized when iteration 21 was taken place, while the minimum error occurred with iteration 26.

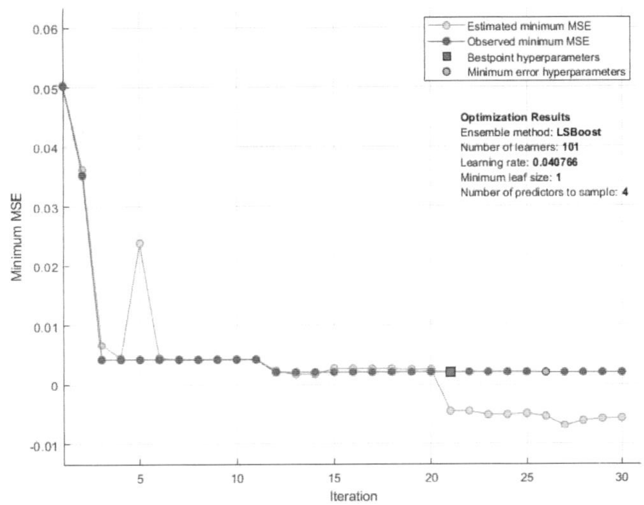

Figure 7: Bayesian hyperparameter optimization performance for ensemble model of GWP

Table 4 presents the actual and predicted fuel consumption and GWP gas emissions values per hour for randomly selected 10 data points (see supplemental data) integrated with the input values of the six independent variables and ship engine types, which cover all engine types in the sustainable LNGC database (diesel, diesel electric, steam, steam turbine and gas turbine). The box plots of the two response parameters clearly stated that optimized fuel consumption ML model has a great prediction accuracy with a median value of 0.052% locating in a 95% confidence interval of [-1.13%, 1.36%], while the median value and confidential interval for the GWP gas emission model is 1.83% within [-1.56%, 5.47%]. These values highly confirmed the prediction robustness of our presented models for the LNGC sustainable cruising contributors. Besides, the proposed models tend to have better prediction performance for SSD diesel engines than gas turbine and steam turbine engines, and it might attribute to the fewer data points applying GT and ST as their engine type or less accurate main engine CO_2 emission data.

Table 4: Excerpt from the supporting information data table showing the prediction of fuel consumption (FC), and GWP gas emission (GWP) using the optimizable models

#	Name	Built Year	Net Tonnage	Molded Breadth (m)	Liquid tank capacity (m^3)	Engine Type	Power (kw)	Speed (kt)	Log(FC)	Pred. log(FC)	FC Pred. error rate (%)	Log(GWP)	Pred. log(GWP)	GWP Pred. error rate (%)
1	Wood side Chaney	2019	34500	46.40	170107	Diesel	21195	15.3	0.42	0.41	0.84	0.94	0.94	0.14
2	Maran Gas Chios	2019	34500	46.40	173400	Diesel Electric(1)	21195	15.3	0.42	0.42	0.63	0.94	0.94	0.14
3	Gaslog Gladstone	2019	34000	45.80	170618	Diesel Electric(2)	42390	19.5	0.72	0.72	0.03	1.24	1.23	0.59
4	Pan Africa	2019	34498	46.95	170442	Diesel Electric(2)	16338	20.4	0.31	0.31	0.11	0.83	0.83	0.05
5	Coral Encanto	2018	7883	28.00	30000	Diesel Electric(1)	8169	16.0	0.06	0.05	0.28	0.57	0.57	0.11
6	Pacific Breeze	2018	434978	52.00	164700	Steam (2)	30060	20.2	0.58	0.59	-1.53	1.09	1.13	-4.15
7	Seri Camara	2018	40083	48.90	147200	Steam Turbine(1)	21195	19.5	0.61	0.60	0.97	1.12	1.08	4.20
8	Pacific Mimosa	2018	41014	48.94	152612	Gas Turbine(1)	25000	19.0	0.68	0.67	1.36	1.18	1.14	4.16
9	Pan Americas	2018	34498	46.95	174000	Diesel (2)	16338	20.4	0.31	0.31	0.18	0.83	0.83	0.41
10	Oceanic Breeze	2018	41069	48.94	155300	Gas Turbine(1)	18510	19.5	0.55	0.55	-0.20	1.05	1.01	4.35

CONCLUSION

In summary, the manuscript provides a reliable prediction for the fuel consumption and GWP gas emissions of 435 LNG carriers based on a thorough comparison among multiple machine learning algorithms with hyperparameter optimization approaches. Based on the LNG ship emission database, two categories of regression algorithms (*i.e.* ensemble of trees and GPR) with consideration of random search and Bayesian optimization are applied in this study. The prediction accuracy performance values have revealed the advantage of the Gaussian process regression approach without optimization in predicting the two response values. However, integrated the GPR and ensemble models with hyperparameter optimization, the optimizable ensemble models presented most accurate and efficient performance. The final models showed excellent prediction robustness on the major contributors of sustainable LNGC cruising. The outputs can be served as an effective tool to solve the lack of quantification for ship fuel consumption and exhaust gas emissions for CO_2, CH_4, N_2O, and black carbon aerosols.

This work provided a holistic way to determine shipping fuel consumption and emissions via hyperparameter optimized machine learning models. The most statistically significant ship static information, which can be utilized to predict ship sustainability performance of newly built LNG tankers, has been successfully identified. The optimized prediction models for global LNG fleet based on a bottom-up method can effectively solve the lack of shipping fuel quantification issue, and the predicted values are reliable sources to validate the accuracy of the collected fuel consumption and GWP gas emissions data. Further investigations may focus on clustering and classification techniques via unsupervised and semi-supervised ML algorithms to fill in gaps for categorizing emerging sustainable fuels from the perspective of maritime safety and sustainability.

REFERENCES

Bond, T.C., Doherty, S.J., Fahey, D.W., Forster, P.M., Berntsen, T., Deangelo, B.J., Flanner, M.G., Ghan, S., Kärcher, B., Koch, D., Kinne, S., Kondo, Y., Quinn, P.K., Sarofim, M.C., Schultz, M.G., Schulz, M., Venkataraman, C., Zhang, H., Zhang, S., Bellouin, N., Guttikunda, S.K., Hopke, P.K., Jacobson, M.Z., Kaiser, J.W., Klimont, Z., Lohmann, U., Schwarz, J.P., Shindell, D., Storelvmo, T., Warren, S.G., Zender, C.S., 2013. Bounding the role of black carbon in the climate system: A scientific assessment. J. Geophys. Res. Atmos. 118, 5380–5552. https://doi.org/10.1002/jgrd.50171

Comer, B., Olmer, N., Mao, X., Roy, B., Rutherford, D., 2017. Black Carbon Emissions and Fuel Use in 2015. Int. Counc. Clean Transp.

DNV GL, 2019. Maritime Forecast To 2050 118.

EMEP/EEA, 2019. EMEP/EEA air pollutant emission inventory guidebook 2019 [WWW Document]. URL http://efdb.apps.eea.europa.eu/?source=%7B%22query%22%3A%7B%22bool%22%3A%7B%22must%22%3A%5B%7B%22term%22%3A%7B%22Fuel%22%3A%22Marine diesel oil%2Fmarine gas oil (MDO%2FMGO)%22%7D%7D%5D%7D%7D%2C%22display_type%22%3A%22tabular%22%7D

ENTEC, 2010. UK Ship Emissions Inventory Final Report.

EPA, 2000. Analysis of Commercial Marine Vessel Emissions and Fuel Consumption Data,.

Eyring, V., Köhler, H.W., Van Aardenne, J., Lauer, A., 2005. Emissions from international shipping: 1. The last 50 years. J. Geophys. Res. D Atmos. 110, 171–182. https://doi.org/10.1029/2004JD005619

Feurer, M., Hutter, F., 2019. Hyperparameter Optimization. https://doi.org/10.1007/978-3-030-05318-5_1

Fortmann-Roe, S., 2012. Understanding the Bias- Variance Tradeoff. Blog.

Hunsucker, J.T., Przelomski, D., Bashkoff, A., Dixon, J., 2018. Uncertainty Analysis of Methods Used to Measure Ship Fuel Oil Consumption.

International Maritime Organisation, 2016. Resolution MEPC.278(70) - Amendments to MARPOL Annex VI.

Ji, C., El-Halwagi, M.M., 2020. A data-driven study of IMO compliant fuel emissions with consideration of black carbon aerosols. Ocean Eng. 218, 108241. https://doi.org/10.1016/j.oceaneng.2020.108241

Ji, C., Jiao, Z., Yuan, S., El-Halwagi, M.M., Wang, Q., 2021a. Development of novel combustion risk index for flammable liquids based on unsupervised clustering algorithms. J. Loss Prev. Process Ind. https://doi.org/10.1016/j.jlp.2021.104422

Ji, C., Yuan, S., Jiao, Z., Huffman, M., El-Halwagi, M.M., Wang, Q., 2021b. Predicting flammability-leading properties for liquid aerosol safety via machine learning. Process Saf. Environ. Prot. 148, 1357–1366. https://doi.org/10.1016/j.psep.2021.03.012

Jiao, Z., Ji, C., Yuan, S., Zhang, Z., Wang, Q., 2020. Development of machine learning based prediction models for hazardous properties of chemical mixtures. J. Loss Prev. Process Ind. https://doi.org/10.1016/j.jlp.2020.104226

Kwon, Y., Lim, H., Lim, Y., Lee, H., 2019. Implication of activity-based vessel emission to improve regional air inventory in a port area. Atmos. Environ. 203, 262–270. https://doi.org/10.1016/j.atmosenv.2019.01.036

MEPC 72, I., 2018. INITIAL IMO STRATEGY ON REDUCTION OF GHG EMISSIONS FROM SHIPS 304, 1–11.

Moreno-Gutiérrez, J., Calderay, F., Saborido, N., Boile, M., Rodríguez Valero, R., Durán-Grados, V., 2015. Methodologies for estimating shipping emissions and energy consumption: A comparative analysis of current methods. Energy 86, 603–616. https://doi.org/10.1016/j.energy.2015.04.083

Moreno-Gutiérrez, J., Pájaro-Velázquez, E., Amado-Sánchez, Y., Rodríguez-Moreno, R., Calderay-Cayetano, F., Durán-Grados, V., 2019. Comparative analysis between different methods for calculating on-board ship's emissions and energy consumption based on operational data. Sci. Total Environ. 650, 575–584.

https://doi.org/10.1016/j.scitotenv.2018.09.045

Olmer, N., Comer, B., Roy, B., Mao, X., Rutherford, D., 2017. Greenhouse Gas Emissions From Global Shipping, 2013-2015. Int. Counc. Clean Transp. 1–25.

Pinho, L.G.B., Nobre, J.S., Singer, J.M., 2015. Cook's distance for generalized linear mixed models. Comput. Stat. Data Anal. https://doi.org/10.1016/j.csda.2014.08.008

Rakke, S.G., 2016. Ship emissions calculation from AIS-annotated.

Rasmussen, C.E., Williams, C.K.I., 2018. Gaussian Processes for Machine Learning, Gaussian Processes for Machine Learning. https://doi.org/10.7551/mitpress/3206.001.0001

Sabine, C., 2014. Ask the Experts: The IPCC Fifth Assessment Report. Carbon Manag. 5, 17–25. https://doi.org/10.4155/cmt.13.80

Smith, T.W.P., Jalkanen, J.P., Anderson, B.A., Corbett, J.J., Faber, J., Hanayama, S., O'Keeffe, E., Parker, S., Johansson, L., Aldous, L., Raucci, C., Traut, M., Ettinger, S., Nelissen, D., Lee, D.S., Ng, S., Agrawal, A., Winebrake, J.J., Hoen, M., A., 2014. Third IMO Greenhouse Gas Study 2014. Int. Marit. Organ. 327. https://doi.org/10.1007/s10584-013-0912-3

Starcrest Consulting Group, L., 2014. Port of Long Beach Air Emissions Inventory in 2013, Journal of Chemical Information and Modeling. https://doi.org/10.1017/CBO9781107415324.004

Uyanik, T., Arslanoglu, Y., Kalenderli, O., 2019. Ship Fuel Consumption Prediction with Machine Learning, in: 4th International Mediterranean Science and Engineering Congress.

Wan, Z., Zhang, Q., Xu, Z., Chen, J., Wang, Q., 2019. Impact of emission control areas on atmospheric pollutant emissions from major ocean-going ships entering the Shanghai Port, China. Mar. Pollut. Bull. 142, 525–532. https://doi.org/10.1016/j.marpolbul.2019.03.053

Witten, I.H., Frank, E., Hall, M. a., 2011. Data Mining: Practical Machine Learning Tools and Techniques, Third Edition, Annals of Physics. https://doi.org/10.1002/1521-3773(20010316)40:6<9823::AID-ANIE9823>3.3.CO;2-C

Yuan, J., Nian, V., 2018. Ship energy consumption prediction with Gaussian process metamodel. Energy Procedia 152, 655–660. https://doi.org/10.1016/j.egypro.2018.09.226

Yuan, S., Ji, C., Han, H., Sun, Y., Mashuga, C. V., 2020. A review of aerosol flammability and explosion related incidents, standards, studies, and risk analysis. Process Saf. Environ. Prot. 146, 499–514. https://doi.org/10.1016/j.psep.2020.11.032

Yuan, S., Ji, C., Monhollen, A., Kwon, J.S.-I., Mashuga, C., 2019. Experimental and thermodynamic study of aerosol explosions in a 36 L apparatus. Fuel 245, 467–477. https://doi.org/10.1016/j.fuel.2019.02.078

SNAME
THE INTERNATIONAL COMMUNITY
FOR MARITIME AND OCEAN PROFESSIONALS

THE ECO POWER SYSTEM – A GWH CLASS UNDERWATER COMPRESSED AIR ENERGY STORAGE SYSTEM

Brian James Cheater
GustoMSC US Inc

Proceedings of the 26th Offshore Symposium, April 6th & 7th 2021, Virtual, Houston, Texas
Texas Section of the Society of Naval Architects and Marine Engineers

ABSTRACT

Energy storage will be essential for grid stability as an increasing amount of intermittent renewable energy (such as offshore wind or solar) is brought on-line. Currently accepted storage solutions are limited by opportunity (pumped hydro) or by limitations on strategic materials such as vanadium, cobalt and lithium. The ECO system does not suffer from these limitations and it is scalable into the gigawatt-hour range using an extrapolation of available offshore technology at a cost low enough to be financially viable using just arbitrage on the daily retail energy market using any combination of wind and solar power. Beyond this, the system has the potential to scale up to provide strategic levels of grid storage for longer averaging periods. This paper will present the technical design of an efficient and viable offshore compressed air energy system.

Keywords: SNAME, Offshore Symposium, CAES, UW-CAES, Energy Storage

Introduction

As the proportion of intermittent renewables increases, the need increases for an energy storage solution that can scale up with the right combination of economic viability, technical assurance and permitting ease.

The problem arises because, without a means of storing electrons, instantaneous production must balance demand on an electrical grid. Consistent baseload capacity is provided by large nuclear or coal fire power plants. This is supplemented by peaker-plants such as gas-fired units which can be throttled up or down to meet the variable demand component.

There are a wide variety of energy storage solutions available with different areas of application as shown in the figure 1 below.

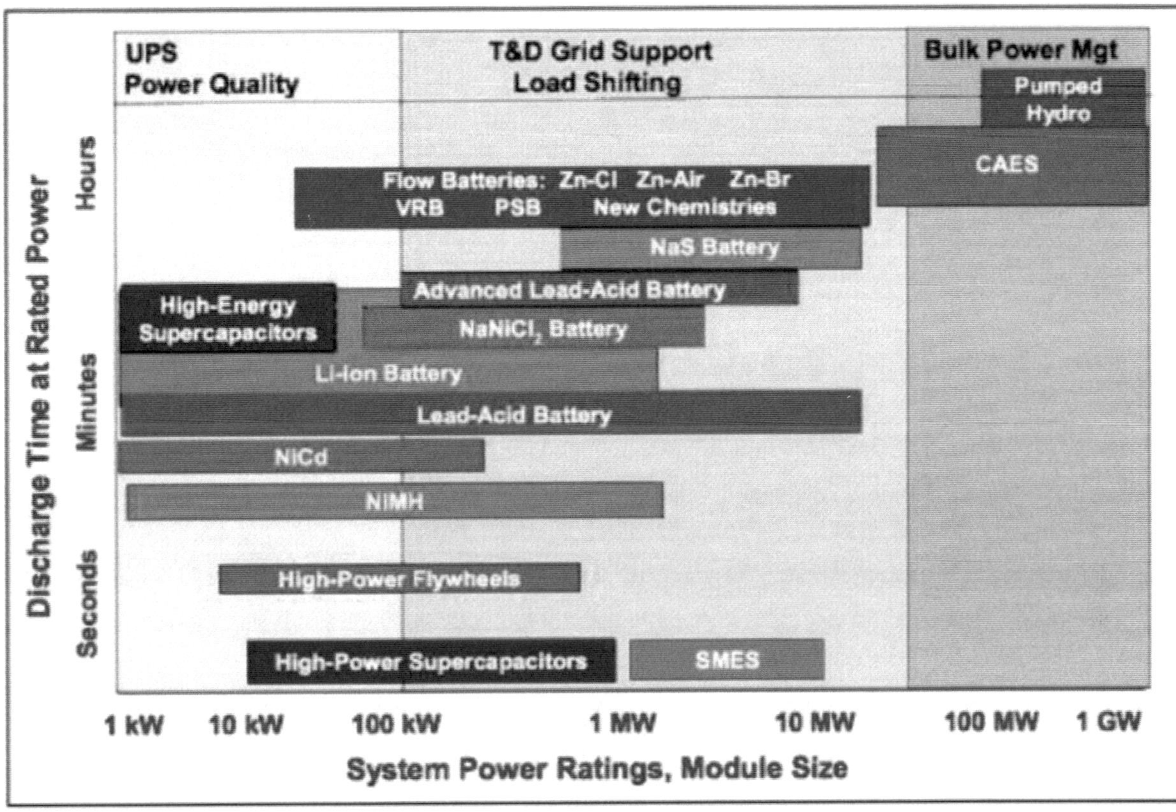

Figure 1: Storage Solutions and Range of Applicability (Ford)

Ultra-fast systems such as super-capacitors or flywheels are low capacity systems intended for frequency regulation or voltage drooping. Fast response systems like batteries or dropped weights can be used for longer duration demand spikes. Prompt-response systems such as flow batteries, pumped-hydro or compressed air can be used in lieu of traditional gas-fired peaker-plants for daily variations.

The proposed system is designed to serve the daily peak shaving market. It is assumed to charge from a regional group of wind farms or solar arrays (see Figure 2) over an 18 hour period and discharge over 6 hours. Beyond peak-shaving, periods of curtailment also open the possibility of power that can be purchased at even lower prices. This is seen as particularly relevant for California which has had to curtail large amounts of un-needed solar capacity.

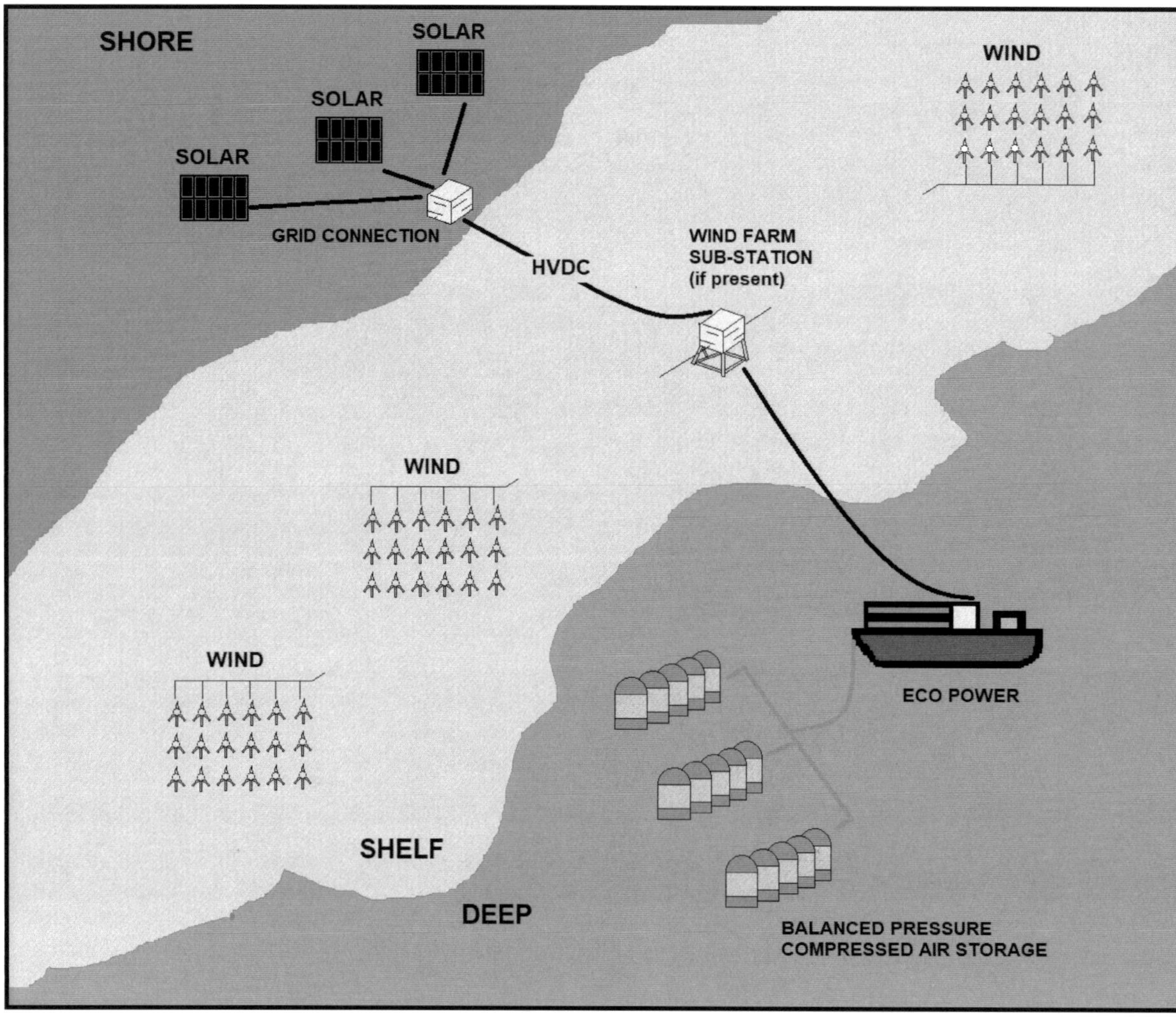

Figure 2: ECO Power System Regional Inter-Connections

SYSTEM DESCRIPTION

The Electrical Compression Offshore (ECO) Power System is classified as an adiabatic underwater compressed air energy storage system (UW-CAES). The system uses an isobaric storage system compared to cavern based systems which are isochoric.

The field layout for the EXO System is shown in Figure 3 below.

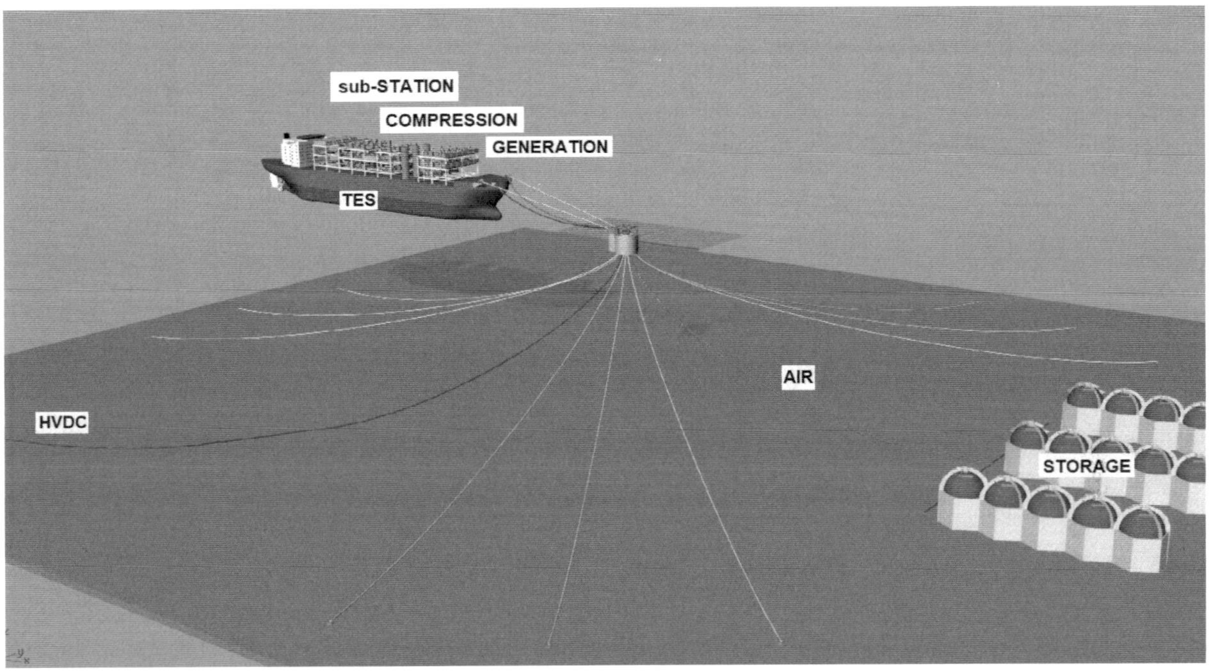

Figure 3: ECO System Field Layout

The ECO System is a compression/expansion plant mounted on a converted tanker which hosts the thermal energy storage (TES) system in its cargo tanks. The ship general arrangement is shown in Figure 4 below.

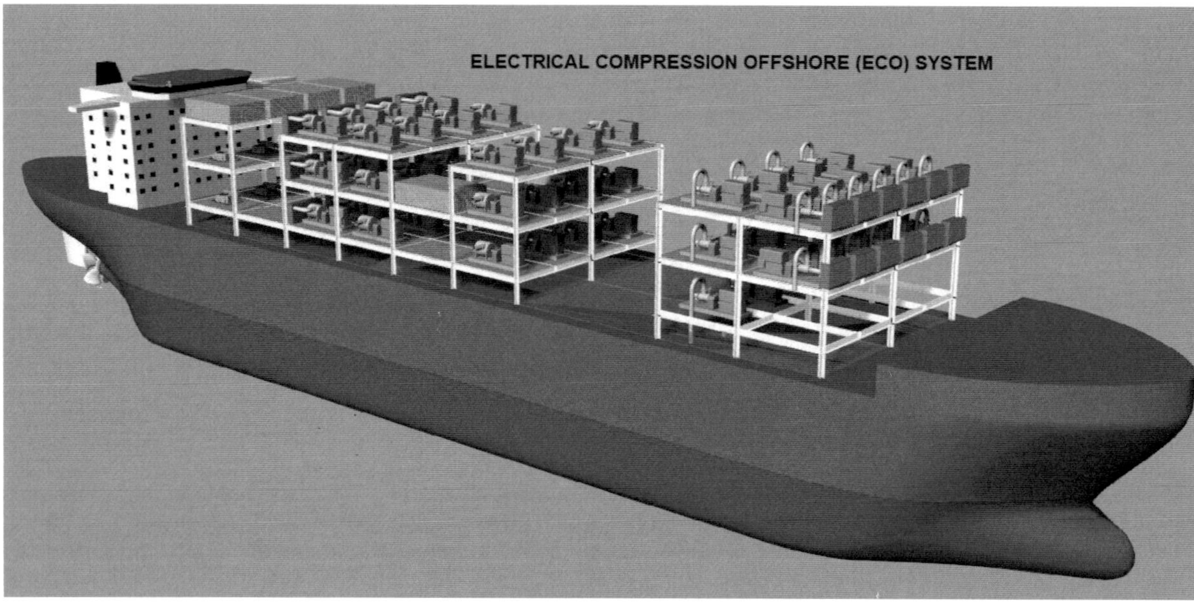

Figure 4: ECO System General Arrangement

When demand is relatively low (e.g. off-peak) or supply is relatively high (e.g. curtailment), power is taken from a regional group of wind farms or solar arrays and used to compress air. Inter-coolers capture the waste heat and store it in hot water kept insulated in cargo tanks. The compressed and cooled air is dried and sent through a riser to a sub-sea storage system located on the seabed. The storage membranes are flexible so they are always pressure balanced against ambient hydrostatic head so the system closely approaches ideal isobaric conditions. When power is required during peak periods or shortage periods, the process is reversed and air is allowed to pass through expanders to generate electricity and inter-stage reheating is applied to recover heat from the TES. The system schematic is shown in Figure 5 below.

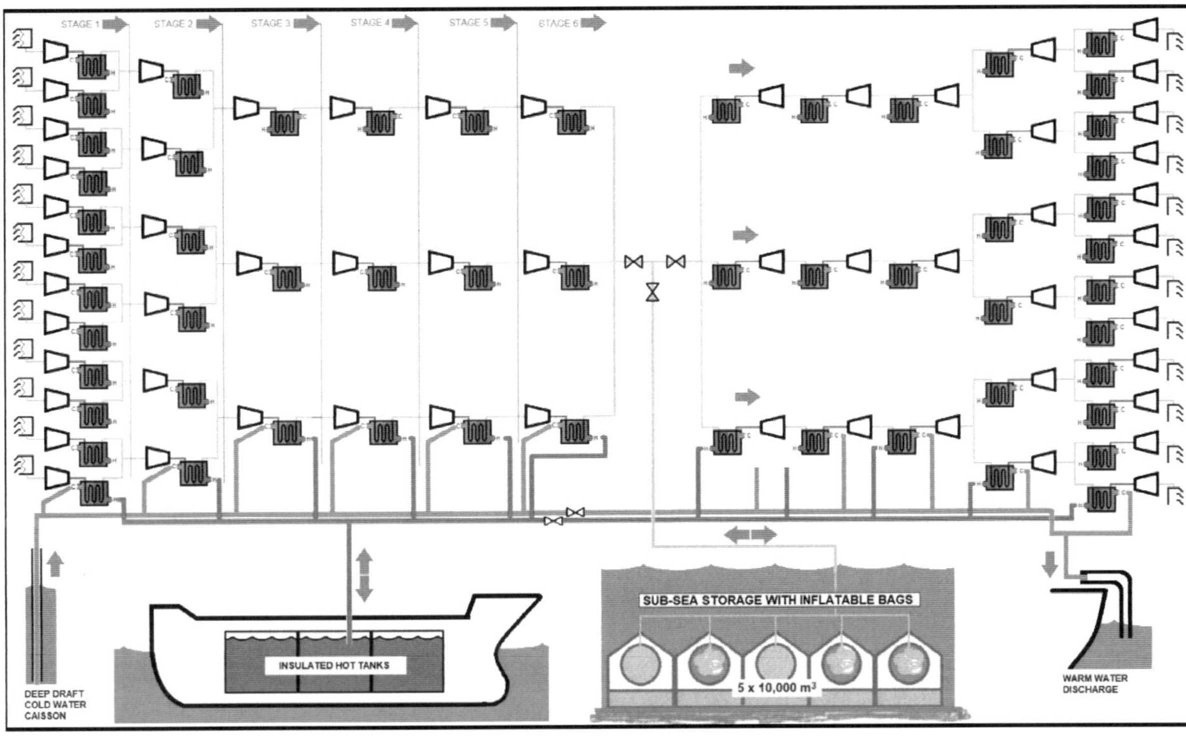

Figure 5: Preliminary System Schematic

STORAGE SYSTEM

The following analysis is based on one (1) bank of storage consisting of five (5) bags illustrated in Figure 6 below.

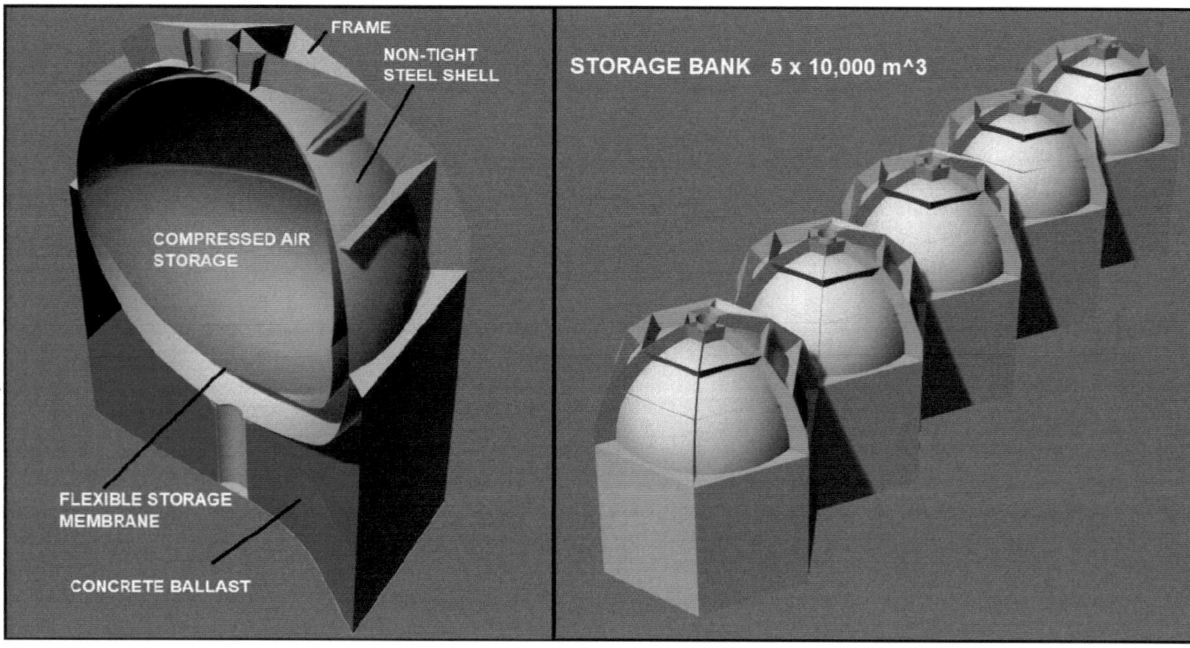

Figure 6: Sub-sea storage system

Each bag is 10,000m^3 leading to a total storage capacity of 50,000m^3.

For preliminary concept analysis, the storage banks were assumed to operate in water depths ranging from 2000m to 6000m but the surface unit and risers will face limitations on what is currently achievable. Duggal et al examined the feasibility of mooring an FPSO in water in waters up to 4500m. For our purposes, we will limit the water depth to 4000m as a reasonable extrapolation of what should be possible. Results of the analysis for the 4000m case are presented below in detail.

Riser options remain to be fully explored but preliminary work indicates a free-standing hybrid riser anchored to the top of the storage bank may be of the most interest. This would be a vertical pipe of X60 high tensile steel supported by a buoyancy can at the surface. The compressed air inside the riser is relatively buoyant and this reduces the net vertical load. Flexible jumpers would be provided at the base and the top for connection purposes.

Assuming a density (ρ) of 1025 kg/m^3 for seawater, the storage pressure (P) at 4000m is 40.221 MPa (397 bar) at an ambient temperature of 4 deg C. At depth, the compressibility factor (Z) is 1.18 and the gas constant (R) for air is 0.287 kJ/kg-K.

$$PV = ZmR_{air}T \qquad (1)$$

Using the ideal gas law (Eqtn 1), the mass of air stored is 21,340te at a density of 426.8 kg/m^3. This corresponds to a system intake of 19,847,520 m^3 of air from the atmosphere at standard temperature and pressure (STP).

The density of air is low enough that significant upward buoyancy loads are generated which must be resisted by the addition of ballast. The amount of ballast varies with depth. At 1000m, 11292te is required while at 4000m, 5432te is required per vault.

The density of the compressed air at these depths is, however, high enough to create a significant back-pressure in the column of air leading up to the surface.

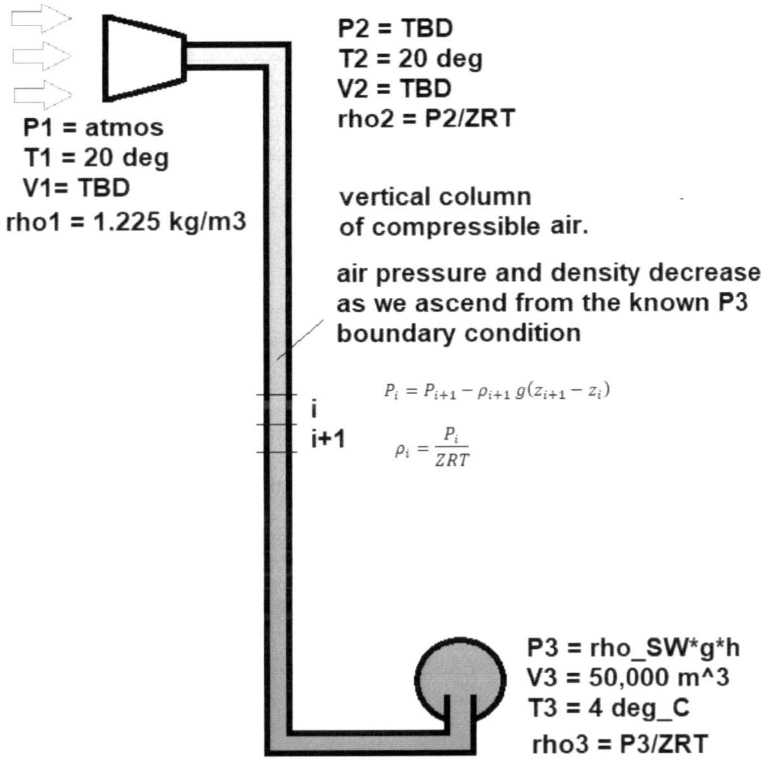

Figure 7: Air Pressure Variation in Riser

Using an iterative back-stepping calculation that accounts for step-wise changes in depth, temperature and compressibility factor (Z), the effective surface pressure which the compression plant must deliver was calculated as 253 bar with a density of 293 kg/m^3 and a mass flow-rate during charging is 426.8 kg/sec.

Further work is required on the mooring and riser design to design and qualify the technology for these depths.

THERMODYNAMIC MODEL

The thermodynamic system is defined as the volume of air sucked into the inlets, compressed and stored in the bags.

As described by Houssainy et al, the general mass and energy balance equations, can be described using the following equations:

$$\frac{dM}{dt} = \dot{m}_{in} - \dot{m}_{out} \tag{2}$$

$$\frac{dE}{dt} = -P\frac{dV}{dt} + \dot{m}_{in}\dot{h}_{in} - \dot{m}_{out}\dot{h}_{out} + \dot{Q} \tag{3}$$

Where M is the mass of air in the system (kg), E is system energy (kJ), h is specific enthalpy (kJ/kg) P is pressure (kN/m^2), V is volume (m^3). The PdV term (kJ) is work done by the system. Q is external heat added to the system (kJ).

In this system, work PdV is done during compression and expansion of the air and during expansion/compression of the sub-sea storage bag.

Mass in equals mass out so the change in energy is due to changes in the enthalpy of the air stream and that depends on the efficiency of the thermal energy storage system (TES)

Q is the external heat added to or removed from the air stream. Heat (Q) is rejected during compression and re-absorbed during expansion. Additional "free" heat input can be drawn from the ocean by allowing the exhaust from the turbo-expanders to drop below zero.

These thermodynamic components of the total energy/work balance are discussed below.

WORK DONE DURING COMPRESSION

The compression will be a multi-stage adiabatic process with inter-cooling. It will be modelled as a polytropic process with a heat capacity ratio (k) of 1.4 as described in the equations below.

$$W_{12} = \left(\frac{k}{k-1}\right) P_1 V_1 \left[\left(\frac{P_2}{P_1}\right)^{(k-1)/k} - 1\right] = \left(\frac{k}{k-1}\right) R_{air} T_1 \left[\frac{P_2}{P_1}^{k/(k-1)} - 1\right] \tag{4}$$

$$P_2 = P_1 \left(\frac{T_2}{T_1}\right)^{\frac{k}{k-1}} \tag{5}$$

$$T_2 = T_1 \left(\frac{P_2}{P_1}\right)^{\frac{k-1}{k}} \tag{6}$$

$$V_2 = V_1 \left(\frac{P_1}{P_2}\right)^{\frac{1}{k}} \tag{7}$$

Because the storage is isobaric and the flow-rate can be controlled, the system will operate under near steady-state conditions so the system can be highly optimized for peak efficiency. Consequently, the compressor efficiency was assumed to be 90%. The number of stages was optimized to approximate the isothermal ideal and yield a compression ratio (r) within equipment limitations. It was found that a 6-stage compression plant led to the highest overall efficiency. The number of turbines at each stage was adjusted to keep the flow rate through each turbine within equipment specifications.

The intercoolers were modeled as isobaric cooling. The heat transfer medium is taken as seawater and the maximum temperature is limited to 100°C so it is single liquid phase with sensible heat only so the enthalpy change is described using:

$$q_c = c_p m (T_2 - T_1) \tag{8}$$

The inter-cooler efficiency is taken as 85%.

The results are summarized in Table 1 and Figure 8 below.

	Inter-cooler Efficiency		**85%**										
	Compressor Efficiency		**90%**										
	Comp Ratio 1st stage		**2.7**										
		n_stage =	**6**	COMPRESSOR								INTERCOOLER	
n	Desc	n_comp	TYPE	r	P1	P2	V1	V2	T1	T2	Spec Wk	qc	V2'
				P2/P1	bar	bar	m^3	m^3	deg_K	deg_K	kJ/kg	kJ/kg	m^3
1	stage	12	Solar C85	2.7	1.0	2.7	18,445,542	9,073,492	283	376	93.4	-93.2	6,831,682
2	stage	6	Solar C85	2.5	2.7	6.7	6,831,682	3,570,802	283	367	84.3	-84.1	2,754,616
3	stage	5	Solar C45	2.5	6.7	16.6	2,754,616	1,439,790	283	367	84.3	-84.1	1,110,694
4	stage	2	Solar C45	2.5	16.6	41.2	1,110,694	580,541	283	367	84.3	-84.1	447,845
5	stage	1	Solar C45	2.5	41.2	102.1	447,845	234,081	283	367	84.3	-84.1	180,577
6	stage	1	Solar C45	2.5	102.1	**253.3**	180,577	94,384	283	367	84.3	-84.1	**72,811**
7													
8													
9													
10													
		27			TGT	**253.34**		**72,811**	Theoretical=		514.8	-513.9	kJ/kg
										Actual=	572.0	-436.8	kJ/kg

Work put into Compression = 1.2E+10 kJ

Heat Available for Storage = 9.3E+09 kJ

Table 1: Compression Process Model

The total work required to compress one bank of air is 1.2E10 kJ or 3390 MWh which represents a power consumption of 188 MW over the 18 hour charging process.

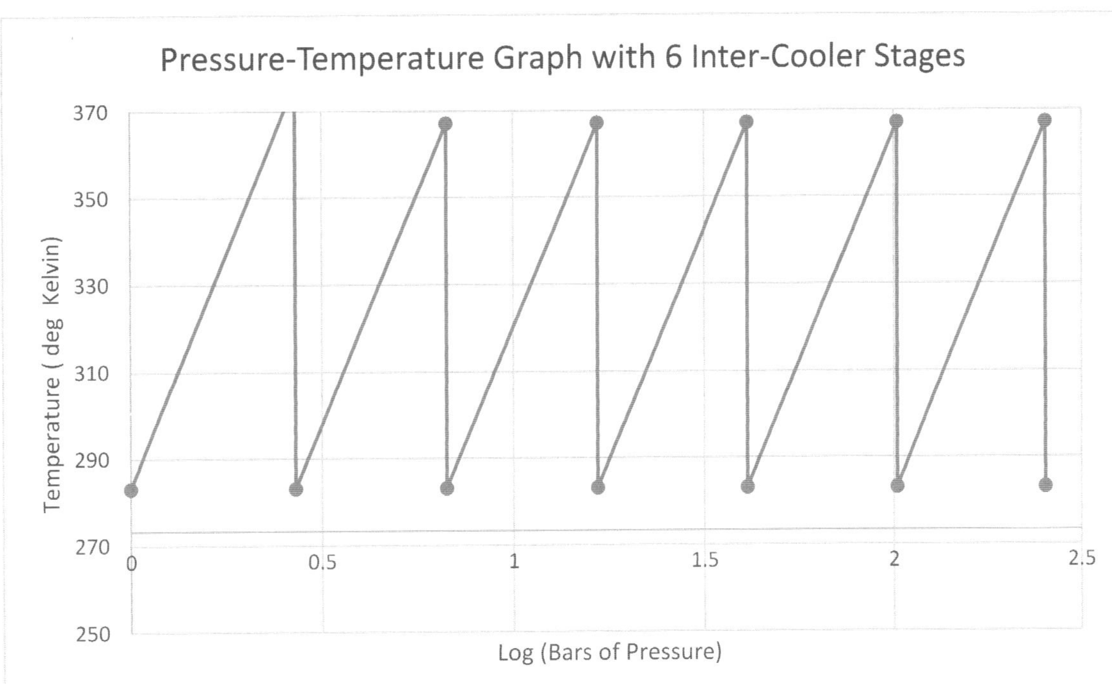

Figure 8: Pressure Temperature Graph during Compression

Further work is required to optimize the compression plant layout and equipment selections.

The system will require support for compressor and particularly turbo-expander manufacturers who would also have to optimize their equipment for a change in medium from natural gas to air and optimize the compression/expansion plants for such large volumes. For example, on stage 1 and 2, the volumetric flow rates are quite high but the pressure and fluid density are low so a larger number of smaller compressors may work better.

However, because the system operates in near steady-state conditions, it should be possible to achieve very high efficiencies which could make these investments worthwhile.

THERMAL ENERGY STORAGE (TES)

Many options exist for thermal energy storage. Thermal oil offers operating temperatures up to 390°C providing high quality heat but it comes at a cost, would have to be replace periodically, and poses a pollution risk. Solar salt (60% NaNo3/40%KNO3) is a non-eutectic mixture with a maximum operating temperature of 550°C which is even higher quality heat with greater exergy. The phase change nature of the material means we get heat storage from both latent and sensible heat modes. However, just like oil, it comes with a cost and would have to be replaced periodically and the material is a pollution risk. In addition to that, the liquidus of 240°C means the salt bath temperature would have to be maintained at an elevated level. Further options exist such as hot-rocks or thermal bricks which may be of interest but all have a cost and were not considered at this preliminary stage.

For this application, seawater was taken as the thermal medium because it is readily available, has a high specific heat and it can be drawn from and returned to the surrounding ocean which acts as both a heat source and a heat sink. Using seawater however, will result in a maximum operating temperature of 100°C which lowers the efficiency of the heat transfer processes which will result in larger heat exchangers and greater pumping loads which will have to be accounted for.

The specific heat (Cp) of saltwater is 3.87 KJ/kg-K. Assuming that water is drawn from the surface at 10°C and heated to 100°C, heat capacity of the seawater storage is 348.3 kJ/kg.

Given a storage requirement of 9.3E9 kJ, the required mass of water for one bank of storage is 26,765te.

The water will be stored in insulated tanks and only for a limited time but for design purposes, a 2% heat loss is assumed leaving 9.13E9 kJ of recoverable heat at a temperature of 98.2°C

Further work is required to design and size the heat exchangers and to investigate the full range of TES options.

Work Done against Gravity

This is the work done lower and raising the mass of air against gravity.

$$W_{12} = mgh \tag{9}$$

Where:

m	=	2.134E7 kg	
g	=	9.82	m/sec^2
h	=	4000	m

W_{12} = 8.374E8 kJ = 232.6 MWh.

It should be noted this will cancel out on a round trip so the work done by gravity as the air drops 4000m to the seabed is reverse by the work that must be done to push the air up 4000m during discharge.

Work Done by Bags to Displace Ocean

This is work done by the boundary of the bag pushing the ambient ocean out during the expansion process.

$$W_{12} = \int P\, dV = P(V2 - V1) \qquad (10$$

Where:

P	=	40,221,000	N/m^2 (external ambient hydrostatic pressure)
V1	=	0	m^3 (empty to start)
V2	=	50,000	m^3 (fully inflated)

W_{12} = 2.011E9 kJ = 558.6 MWh.

It should be noted that storage membrane derived from NOV's Sub-sea storage system is a flexible membrane but is not intended to be elastic or stretch; as it is inflated, it will unfold and as it is deflated it will fold in on itself.

The PV work done to move the membrane boundary out against the ocean is 100% recoverable during discharge.

This stands in contrast to an isochoric system where the fixed boundary means that pressure and temperature drop as the system is discharged thus requiring an external injection of heat to completely recover the energy.

Work Done During Expansion

The expansion process is an adiabatic polytropic process similar to compression.

The discharge time was 6 hours resulting in a discharge flow-rate of 987.6 kg/s.

The efficiency of the heat exchangers was taken as 85%.

Turbo expander efficiency can vary depending on the deviation from the specific speed (see Figure 9).

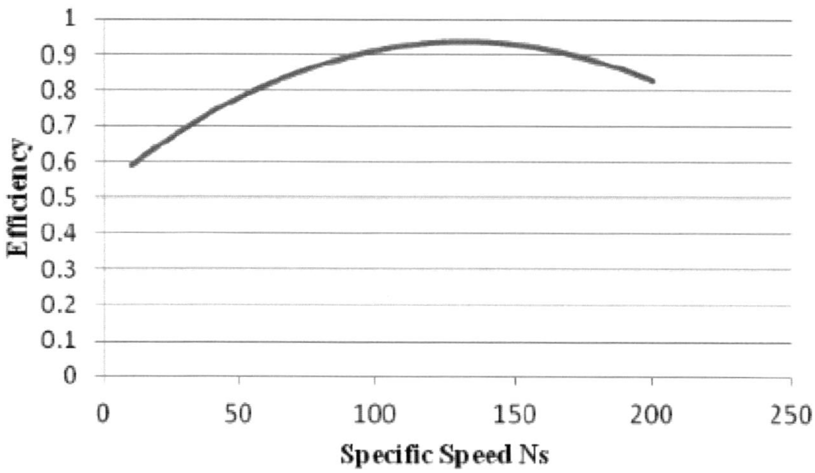

Figure 9: Radial Turbo Expander Efficiency as a Function of Specific Speed (Rezaie et al)

In a natural gas applications, the flow rate varies so IPIECA report effective turbo-expander efficiencies of 80% to 90%. However, in this application, the isobaric nature of the storage means the flow-rate will be very uniform so the turbo-expander will continuously operate very close to its specific speed so the efficiency was taken as 92%.

Re-Heater Eff = **85%**
Expander Eff = **92%**

n_stage =		5	OCEAN preHEAT		Off/On (0/1) :		1	HEAT DEMANDED BY AIR FROM TES PER STAGE					TURBO-EXPANDER							
n Descri	ranche	#expand	qh1	T1	T2	V1	V2	Q23	'3 @85	T2	T3	V3	r	P3	P4	V3	V4	T3	T4	Spec Wk
			kJ/kg	deg_K	deg_K		m^3	kJ	deg_K	deg_K	deg_K	m^3	P2/P1	bar	bar	m^3	m^3	deg_K	deg_K	kJ/kg
1 stage	1	3	6.0	277.15	283.15	**72,811**	74,387	1.83E+09	73	283	356	93449	3.0	**253.3**	83.7	93,449	206,045	356	259	96.7
2 stage	1	3	24.0	259.26	283.15	206,045	225,028	1.83E+09	73	283	356	282694	3.0	83.7	27.7	282,694	623,306	356	259	96.7
3 stage	1	6	24.0	259.26	283.15	623,306	680,730	1.83E+09	73	283	356	855176	3.0	27.7	9.2	855,176	1,885,562	356	259	96.7
4 stage	1	6	24.0	259.26	283.15	1,885,562	2,059,275	1.83E+09	73	283	356	2586991	3.0	9.2	3.0	2,586,991	5,704,008	356	259	96.7
5 stage	1	12	24.0	259.26	283.15	5,704,008	6,229,508	1.83E+09	73	283	356	7825899	3.0	3.0	1.0	7,825,899	17,255,179	356	259	96.7
6																				
7																				
8																				
9																				
10																				
		30	101.8	kJ/kg				9.14E+09						TGT	**1**		17,255,179	m^3	Theoretical	483.7
		Efficien	100%	unlimited supply of heat, just size exchangers										Corrected to Ambient Temp:			18,844,868	m^3	Eff	0.92
		Actual	101.8	kJ/kg															Actual	445.0
			2.17E+09	kJ				OK, NOT CRYOGENIC												9.50E+09
T_out : -13.9 deg_C			604	MWh recovered from ocean													Power Output (MWh) =			2638

Table 2: Expansion Process Model

The expansion process will result in the air-stream cooling which can be used to increase the operating temperature range by going below the temperature of the ambient ocean which can then be used as a heat source. Using an expansion ratio of 3.0 will result in an outlet temperature of -14°C (259K). This is warm enough that it is not cryogenic or require exotic piping material. But, it is cold enough that energy can be absorbed from the ocean in a pre-heating stage. The ocean has the advantage that it is, for practical purposes, an unlimited supply of constant temperature heat energy so as much heat as needed can be extracted. Ocean preheat increases the overall efficiency from 67% to 75.2%.

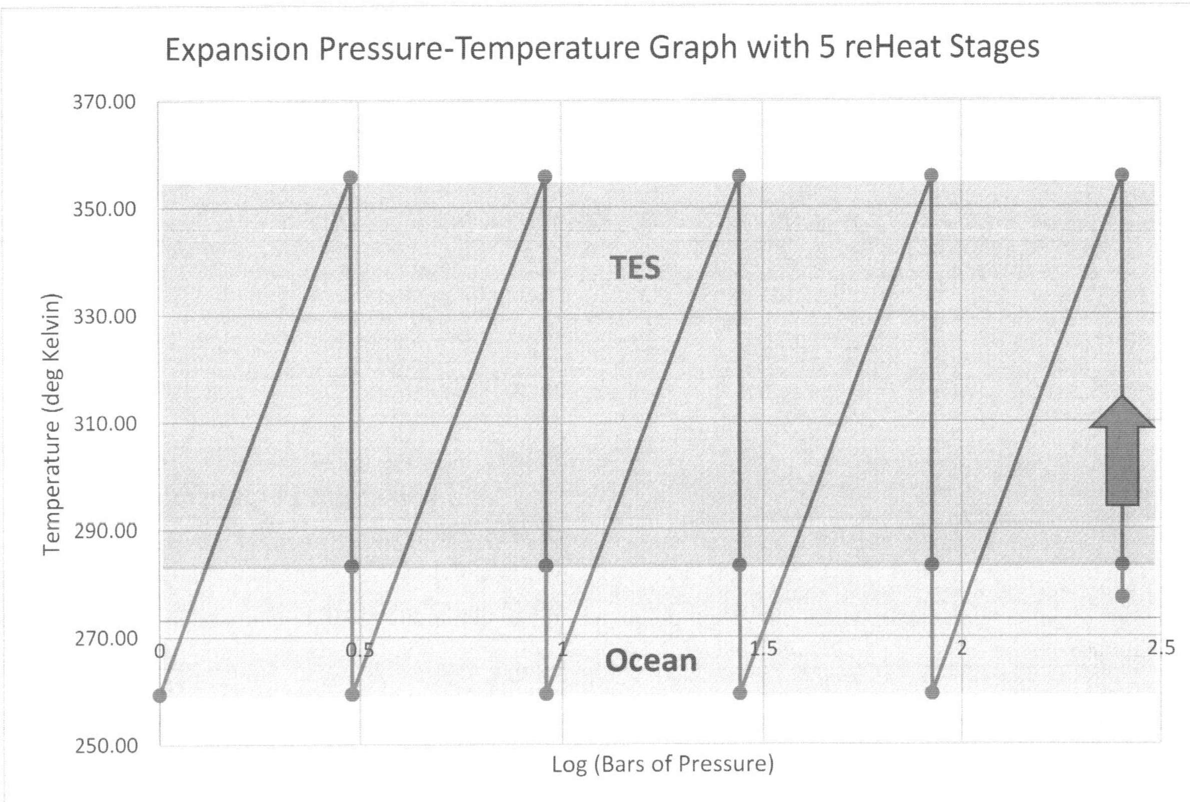

Figure 10: Pressure Temperature Graph during Expansion

Increasing the pressure ratio on the turbo expanders would increase the amount of ocean preheat and increase the operating temperature range but could pose challenges to the design of the turbo expanders and the heat exchangers where low temperature steel and ice formation would become problems with decreased temperature. Additional work is required to optimize the system with respect to heat exchanger size, optimal expander output temperature and number of stages as well as the low temperature heat exchangers.

WORK LOST TO PUMPING, FRICTION AND AUXILIARIES

A riser diameter of 26 inches is sufficient to achieve the required discharge velocity of 2.3 m/s from the storage system.

Assuming a surface roughness of 0.1mm, the work done to overcome friction in the pipeline is 0.1 MJ per discharge. The charging cycle operates at a lower fluid speed and is expected to have lower friction losses but to be conservative, this shall be applied to the charging cycle as well.

Each heat exchanger is assumed to require its own pump which for preliminary assessment purposes is assumed to have an average head of 20.5m stretching from mid-height to cargo hold up to the main deck plus 3m airgap + 5m into the topsides modules. The flow-rate is taken as 1360 m^3/hr.

Assuming 85% efficiency on pumping, the power for an individual pump was calculated as 80kW.

For the charge cycle, the total work lost to pumping was estimated as 44 MWh and for discharge cycle, the total work lost to pumping was estimated as 80 MWh.

The loads due to hotel and auxiliary services was taken as 5% of the pump loads and losses dues to electrical conversion and transformers were taken as 1% of the total load consumed or regenerated.

WORK BALANCE SUMMARY AND SYSTEM EFFICIENCY

The overall work balance summary is shown in Tables 3 and 4.

TOTAL WORK DONE TO FILL STORAGE BAGS (POWER DEMAND)			
	Energy		
Demand Component	kJ	MWh	Comments
Air Compression	1.22E+10	3391	work done on air by compressors
Utility Loads	1.52E+08	44	additional work due to mechanical friction, pumps, support services, hotel loads
Work Against Gravity	-8.37E+08	-233	pushing air down hill, helps to reduce power demand
Bag Inflation Work	2.01E+09	559	inflation work done against ambient hydrostatic pressure
1% Electrical Losses	1.35E+08	38	
Power Absorbed	**1.37E+10**	**3799**	211 MW of power absorbed for 18 hours

Table 3: Compression Work Summary

TOTAL WORK AVAILABLE TO REGENERATE ELECTRICITY (POWER AVAILABLE)			
	Energy		
Demand Component	kJ	MWh	Comments
Air Expansion	9.50E+09	2638	work done by air expansion in turbo-expanders
Utility Loads	-2.73E+08	-80	additional work due to mechanical friction, pumps, support services, hotel loads
Work against Gravity	-8.37E+08	-233	pushing air up hill, reduces power available
Bag Deflation Work	2.01E+09	559	deflation work done by ambient hydrostatic pressure
1% Electrical Losses	-1.04E+08	-29	
Power Recovered	**1.03E+10**	**2855**	476 MW of power produced for 6 hours

Table 4: Expansion Work Summary

The overall efficiency is 75.2% defined as

$$Efficiency\ (\eta) = \frac{Total\ Energy\ Out}{Total\ Energy\ In} \tag{11}$$

Elemgaard et al report an expected efficiency range of 70-80% for adiabatic CAES systems and this is in line with that.

TOTAL SYSTEM STORAGE CAPACITY

Only 26,765te of water needs to be stored for one bank of storage so there is room for additional storage aboard a single ECO Power system.

Assuming 5% of cargo hold volume lost to insulation, 15% for heat exchanger volume and adding 5% margin to the required storage the possible number of storage banks according to vessel size is shown in Table 5 below.

TYPE	L	DWT_LO	DWT_HI	Number of Banks		ECO Capacity (kWh)	
	(m)	(te)	(te)	LO	HI	LO	HI
ULCC	415	320,000	550,000	7	14	19,987	39,975
VLCC	330	200,000	320,000	4	7	11,421	19,987
SUEZMAX	275	120,000	200,000	2	4	5,711	11,421
AFRAMAX	240	80,000	120,000	1	2	2,855	5,711
PANAMAX	225	50,000	80,000	0	1	0	2,855
HANDY	220		50,000	0	0	0	0

Table 5: Total System Capacity by Size of Tanker Conversion

Based on expected topsides weight, the upper bound on the number of banks is expected to be four (4) placed on a large SUEZMAX or small VLCC tanker.

Below, in Figure 11, these capacities are super-imposed on a graphic from Nejad et al for evaluation purposes. As we can see, the ECO Power System is capable of providing power storage levels approaching many pumped hydro energy storage (PHES) projects.

The specific storage capacity is 57.1 kWh/m^3.

For comparison purposes, lithium-ion can achieve an energy density of 220 kW/m^3 with a lithium-nickel-manganese-cobalt (NMC) chemistry which makes it well suited for electric vehicle (EV) applications where individual cars like Tesla having battery capacities of 85-100 kWh. But, there are real constraints on the supply of cobalt and to a much lesser extent lithium. Efforts are being made to increase battery recycling and changing the cathode chemistry to other chemistries like lithium-iron-phosphate but it is not clear how these will develop or be adopted over time or how they will scale up beyond the electric vehicle industry which by itself will be a large source of demand. Flow batteries are interesting possibilities but it needs to be demonstrated how they scale-up as well.

The hydrogen economy has garnered a lot of attention but significant challenges exist for large scale deployment. Brown hydrogen generation from natural gas is well established and economical but is not carbon free. Green hydrogen is carbon-free but it relies on electrolysis which is relatively expensive. Practical electrolyzers operate around 50 kWh/kg and while improvements are expected, the minimum possible energy input is dictated by the Gibbs energy required to split water molecules which is 39 kWh/kg. In addition to that, while hydrogen (120 MJ/kg) may have greater specific energy density than gasoline (42 MJ/kg), it is a gas at standard temperature and pressure, so the volumetric energy density is very low unless it is highly compressed (300 to 700 bars) or cooled to cryogenic temperatures (-252.87 deg_C). Using caverns for bulk storage of hydrogen would still require a distribution network to reach consumers with many of the above problems. Converting the hydrogen to ammonia does simplify storage and transportation but it adds to the complexity of the supply chain and does not completely solve the problem since ammonia is toxic and it is also a gas that must be pressurized and/or cooled (-33 deg_C) to liquefy it. Furthermore, even though hydrogen fuel cells are relatively efficient (up to 60%) there are still significant losses converting this back to useful electrical energy. Ammonia fuel cells are slightly less efficient (40-50%) and the development of direct ammonia combustion engines is still an ongoing effort with, for example, the first commercial marine engines expected from MAN Diesel in 2024. The efforts at direct ammonia combustion are challenged by the low propagation speed of the flame, corrosiveness, and the production of nitrous oxides (NOx).

Another purely mechanical storage system, pumped hydro, has enjoyed considerable success but is heavily constrained by site availability. Competing cavern storage technologies require the availability of caverns and then have efficiency challenges due to the isochoric nature of the storage chamber which results in a dropping pressures/temperatures during discharge which have to be compensated for with external heat. Compressed liquid air is an interesting option but is still emerging. Other concepts such as dropped weights are prompt-response systems with limited capacities intended for other parts of the energy storage market.

These are all areas of active research and the cost of renewable energy (particularly with solar) is dropping extremely quickly so it should be interesting to see what developments do occur but for the moment, it would appear that there may still be opportunity for other storage technologies such as ECO Power to develop.

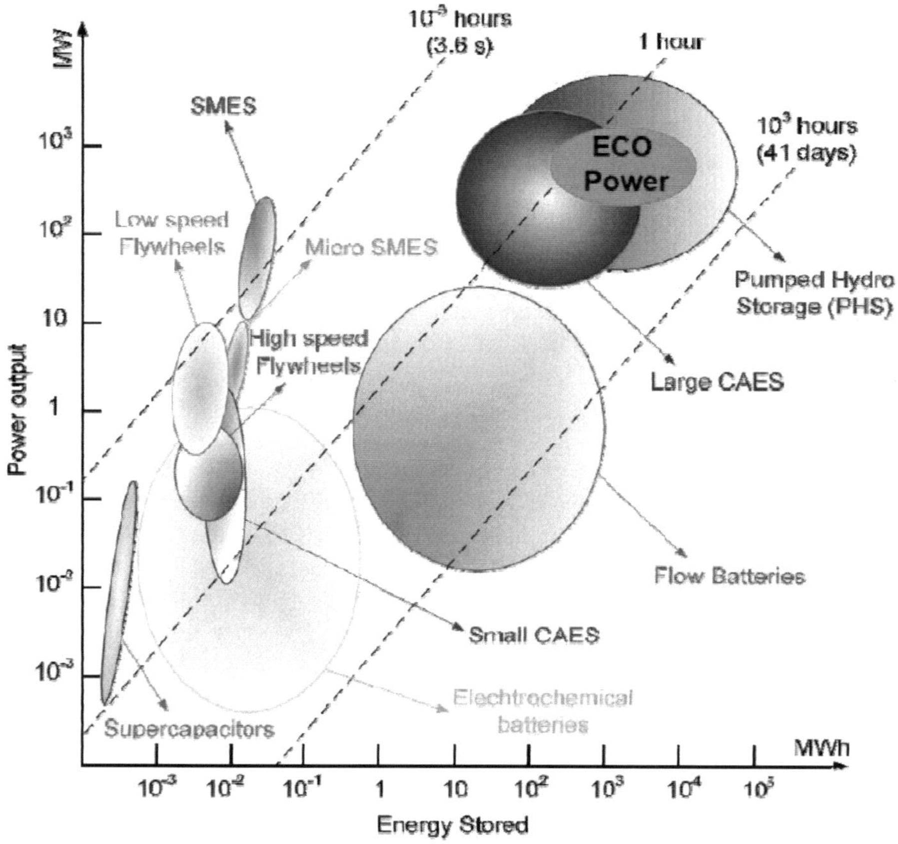

Figure 21: Comparison of Power and Energy Storage Capacities

PRELIMINARY COST ESTIMATE

A large part of the value of an energy storage system is in the resiliency it brings to the grid. Some grid operators (such as IESO in Ontario) offer capacity availability payments and ancillary stand-by capacity or operating reserve payments in addition to potential revenue from market arbitrage. But this can be difficult to quantify and policy is not consistent across all jurisdictions so the preliminary economic assessment presented here is based on project economics around a market arbitrage model and cost comparison (LCOS and $/kWh) to other CAES systems and competing technologies such as batteries.

A preliminary cost estimate was completed. The cost estimate assumed the hull was a tanker conversion. Costs were estimated for compressors, turbo-expanders, steel, piping, heat exchangers, deck reinforcement, insulation, moorings and risers, sub-sea storage, installation and project management. The sub-station costs were based on those found on fixed platforms in wind farms. The HVDC electrical import/export line cost was also estimated but not all competing concepts include this so two sets of cost numbers are quoted – one with and one without HVDC costs.

The levelized cost of storage (LCOS) was calculated using the formula below.

$$LCOS\left(\frac{\$}{kWh}\right) = \frac{\sum_i\left(\frac{CAPEX_i + O\&M_i + CHARGING\_COST_i}{(1+r)^n}\right)}{\sum_i\left(\frac{POWER\_PRODUCED_i}{(1+r)^n}\right)} \qquad (12)$$

A discount rate of 10% and inflation rate of 1% were assumed. The owner's leverage ratio on debt was assumed to be 30% with financing costs of 5% over a project life of 30 years. A bath-tub curve was adopted for the O&M with a 3-year burn-in time, mid-life annual O&M of 0.5% of CAPEX and cost escalation with age to 1.5% of CAPEX annually at end of life.

The $65/MWh price spread for arbitrage was based on an average from the US retail markets with a correction to wholesale pricing. It does not consider the possible benefit from power purchases during curtailment events where low purchase prices might be negotiated for power production potential that would otherwise be lost.

Figures 12 and 13 below show $/MWh and LCOS versus the number of banks both with and without the costs of the HVDC export line included.

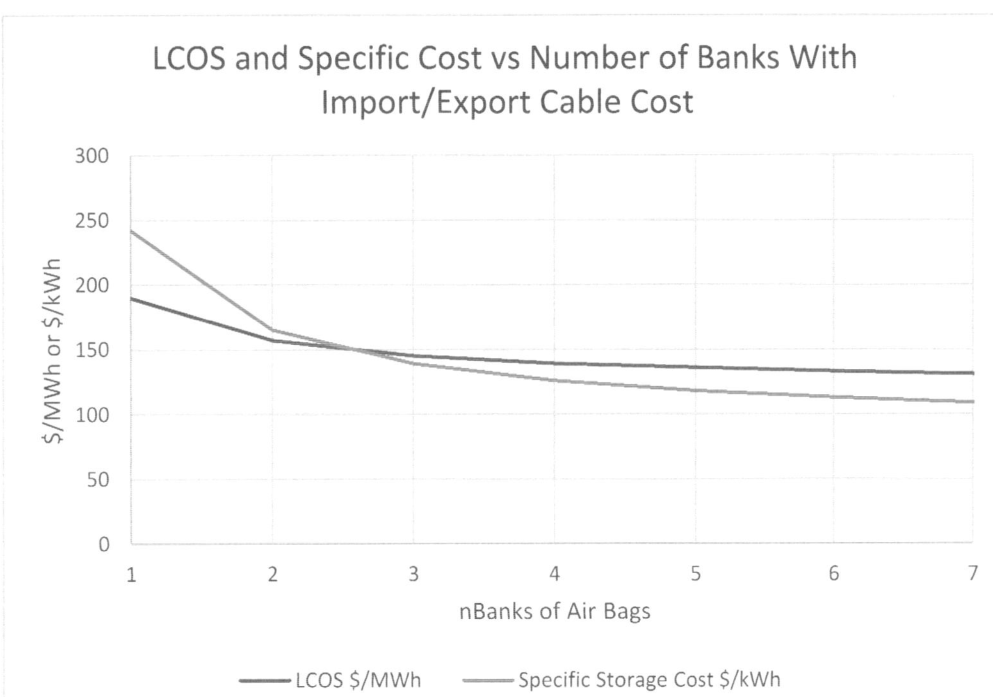

Figure 12: LCOS and $/kWh including Export Cable Cost

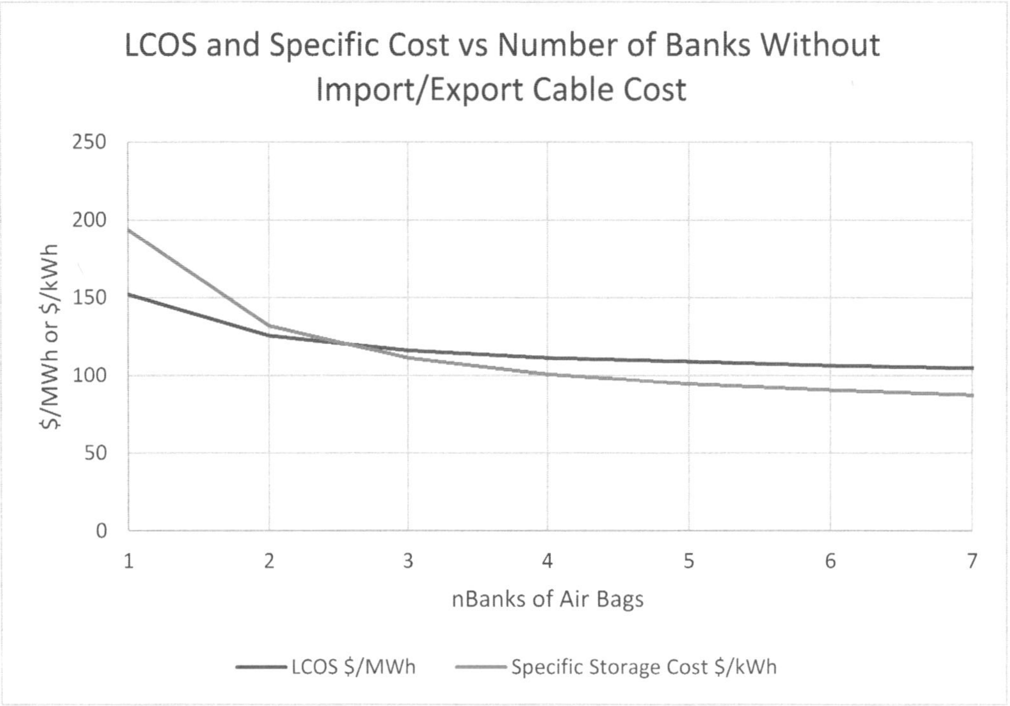

Figure 13: LCOS and $/kWh not including Export Cable Cost

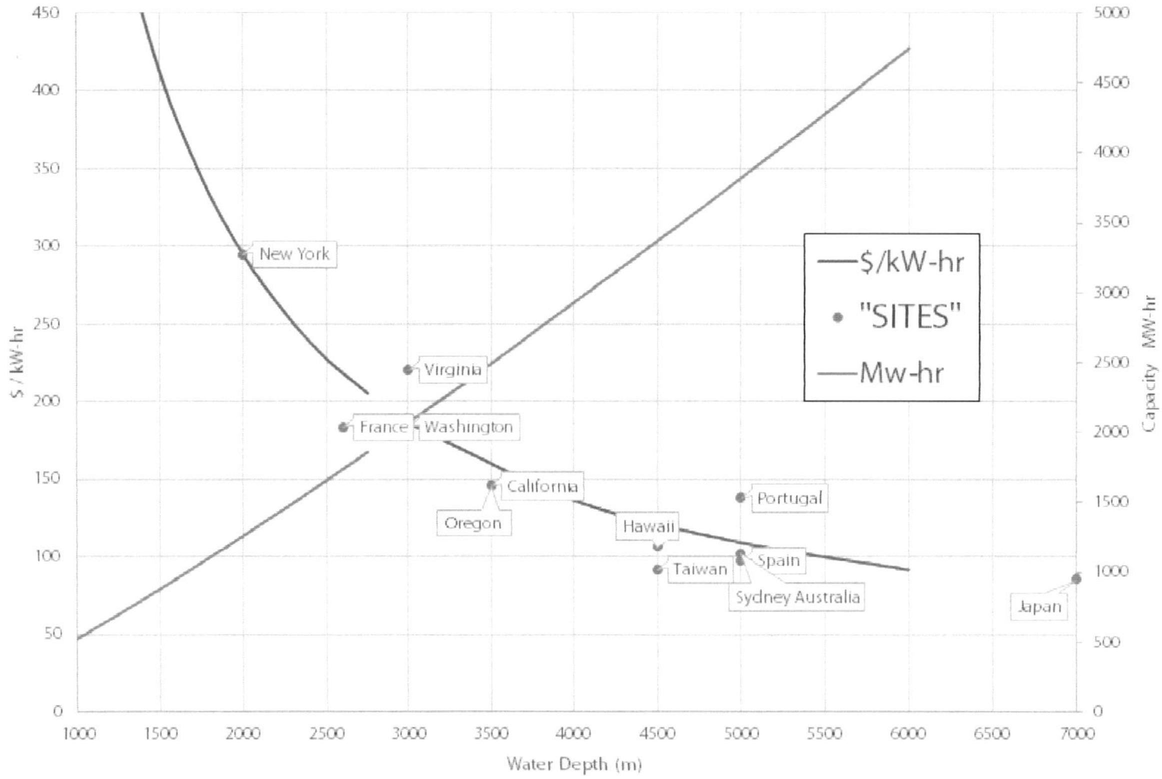

Figure 14: Indicative $/kWh (incl HVDC) for Sites World-Wide

There is substantial drop in cost with the first few increments in the number of banks. This is because the fixed overhead costs due to vessel conversion, HVDC cable, moorings and risers and offshore installation are spread out but this diminishes as the incremental costs from the larger topsides start to make themselves felt.

Figure 14 shows the indicative costs for sites world-wide where deep-water is within 60 miles of the transmission terminus. Deep-water sites (>4000m) in Hawaii to Japan are show for comparison purposes, it is of course possible to install in waters ≤ 4000m in these locations should the deeper water not be attainable.

It is interesting to look at some of the many competing technologies.

The specific energy cost is as low as 83.2 $/kWh which is comparable to 84.31 $/kWh reported by Dodaro for some pumped hydro energy storage (PHES).

This is also less than the stated price goal of 100 $/kWh set for lithium ion batteries. Of course, this must be understood to be a moving target due to continued technology refinements and build-out of the supply chain which could reduce costs even more if there are no material constraints due for example to resource exhaustion or geopolitical constrictions.

Conclusions and Future Work

Preliminary work indicates the ECO Power System may be an interesting and viable UW-CAES solution with a robust underwater storage solution suitable for grid-scale energy storage.

The system does require ultra-deep water but it is believed to be a reasonable extrapolation of existing practice and this does generate the very high storage densities that drive the storage capacity up. It is these economies of scale that make this type of storage system viable in a high overhead environment such as offshore.

But working offshore does eliminate many of the permitting problems, environmental challenges, lobbying efforts from special interest groups and land acquisition costs associated with onshore development which are not trivial barriers in a lot of cases.

Nothing in this concept is fundamentally new, but it is an ambitious extrapolation of currently existing technology that will require technology development and qualification. Such an extrapolation is necessary to achieve the economies of scale that make the concept viable but it does mean development efforts must be approached in a careful and systematic way.

One possible development roadmap is outlined in the steps below.

1) Create a group of industry champions and appropriate discipline experts. Industry champions could be from a grid operator, regulator or government policy expert, offshore wind developer or energy company. Discipline expertise would have to include: compressors, turbo-expanders, heat exchanger, sub-sea storage engineering, construction and installation, moorings, risers, naval architecture, utility-scale power generation, grid modelling and inter-connection, grid policy and energy/carbon price modelling. For initial purposes, only a core group of essential technologies may be required.

2) Engineer a concept level design. Validate the concept and the project economics. Identify main technical/commercial and strategic/policy challenges (Fig 15)

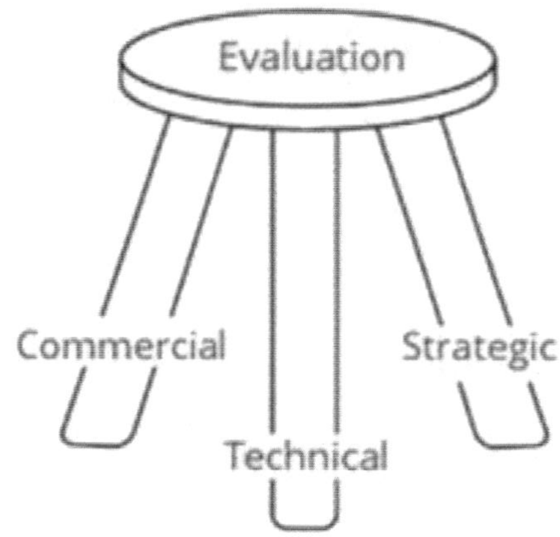

Figure 15: Components for Evaluation

Define grid operator requirements on quality and reliability of electrical regeneration capacity. Assess environmental impact and carbon curtailment potential. Identify market participants and begin out-reach to introduce concept.

3) Create a detailed development plan with clearly defined targets, personnel requirements, schedules and budgets.

4) Identify and secure funding from some appropriate mechanism such as a joint industry project (JIP) for initial stages at least, public grants and investment (DOE, NYSERDA, CAISO) and/or private investment (Oil and Gas Company interested in energy transition, Wind Farm Developer, NGO, Foundation)

5) Perform more detailed engineering, complete with hazard identification, risk assessment and assessment of robustness and resiliency for grid connection. Refine and revalidate concept and economics. Obtain 3rd party certification.

6) Design and Build a proto-type system (e.g. one storage chamber in shallow water hosted off the back of a supply boat or similar) as proof of concept to achieve a technology readiness level (TRL) of 3.

7) Refine design and economics with lessons learned. Revalidate concept and economics. Market validated design.

8) Build a complete operational system connected to the grid.

This is not a quick path to a full solution but, given the scale and complexity of creating new concepts that are validated and bankable, a quick path should perhaps not be expected. This is expected to be a 5-10 year process but this timetable aligns with deployment of multiple regional-scale offshore wind projects on the East Coast of the US. Looking beyond the US, figure 16 highlights the potential growth in the energy storage market. Such a market could support multiple instances of the ECO Power System world-wide which makes this concept worth investigating.

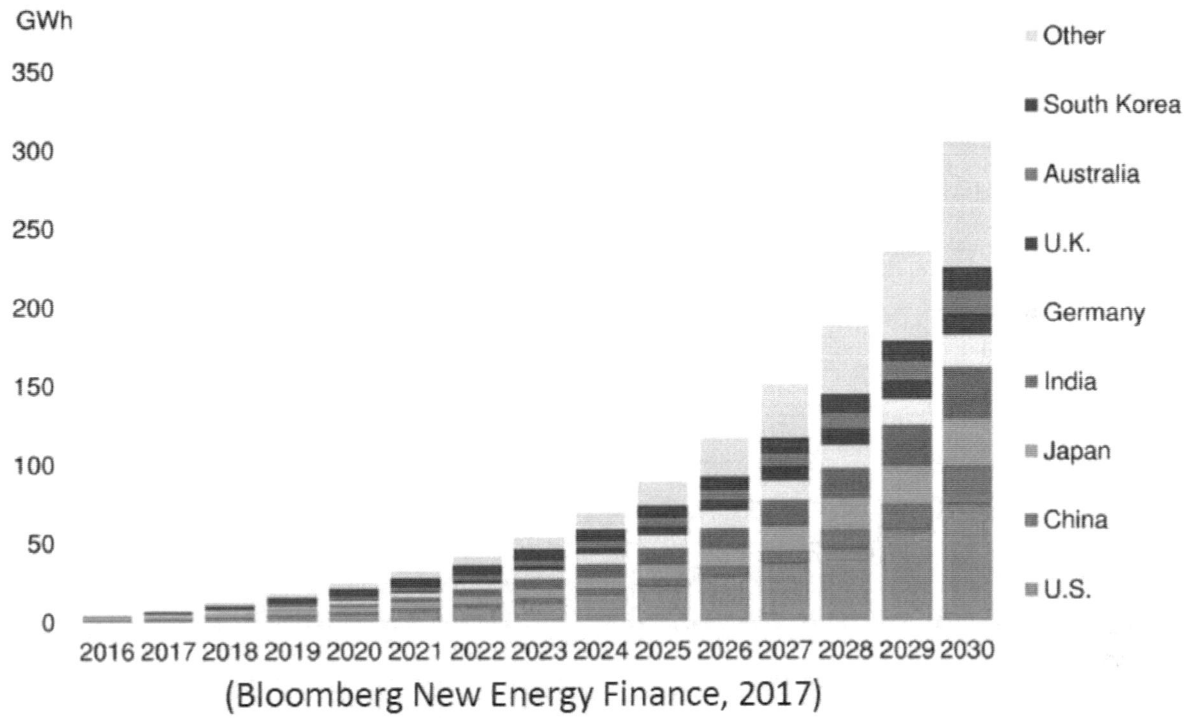

Figure 16: Energy Storage Market Predictions out to 2030 (Bloomberg)

REFERENCES

Bieter, Heeter, Spitsen and Riley (2020) Comparing Offshore Wind Energy Procurement and Project Revenue Sources Across US States. Report NREL/TP-5000-76079 National Renewable Energy Laboratory.

Bloomberg, New Energy Finance, 2017

Dodaro, J. (2015) Molten Salt Storage. http://large.stanford.edu/courses/2015/ph240/dodaro2/

Duggal, Izadparast and Pothuganti. (2015) Stationkeeping Technology for Frontier Deepwater Floating Systems OTC-26058-MS

Elmegaard, Brian and Brix, W. (2011) "Efficiency of Compressed Air Energy Storage", 24th International Conference on Efficiency, Cost, Optimization, Simulation and Environmental Impact of Energy Systems

Ford, A. (2015) Unleashing the Value of Energy Storage: A Case Study of Fuel Free, Compressed Air Energy Storage for the Ontario Power System, NRSTOR.

Houssainy, Janbozorgi and Kavehpour. (2018) Theoretical Performance Limits of an Isobaric Compressed Air Energy Storage System. ASME J. Energy Resources Technology 140(10) p.101201

International Petroleum Industry Environmental Conservation Association (IPIECA) https://www.ipieca.org/resources/energy-efficiency-solutions/power-and-heat-generation/turbo-expanders/

Letcher, T (2016) Storing Energy with Special Reference to Renewable Energy Sources. Elsevier

Nejad, Hanif, Iqbal, Tariq and Quaicoe, John Compressed Air Energy Storage System Control and Performance Assessment Using Energy Harvested Index

Zehtabiyan-Rezaie, Navid & Saffar-Avval, Majid. (2012). Feasibility Study of Turbo Expander Installation in City Gate Station. Proceedings of ECOS 2012

TOWARDS A ROBUST OFFSHORE BONDED REPAIR STRENGTH EVALUATION

Stephane Paboeuf, Luc Mouton
Bureau Veritas Marine & Offshore, Nantes, France

Quentin Sourisseau
Universite Gustave Eiffel, Bouguenais, France

Anne-Charlotte Goupil
Siemens, Chatillon, France

Proceedings of the 26th Offshore Symposium, April 6th & 7th 2021, Virtual, Houston, Texas
Texas Section of the Society of Naval Architects and Marine Engineers

ABSTRACT

Since the early 2000s, the number of Floating Production, Storage and Offloading (FPSO) units is increasing significantly. And so now, half of the fleet is over than 10 years old. As FPSO are mainly installed in tropical areas, with marine environment, high temperature and high humidity, corrosion is a permanent threat. Maintenance of steel structures become a challenge for oil major companies in offshore operation. Indeed, when allowable corrosion limit are reached, plates are to be repaired. However the current "crop and renew" technique implies a number of major issues for owners such as: "hot work", i.e., welding; temporary structure weakening; necessity to empty, clean and vent oil tanks, leading to a long down time and an expensive solution.

"Cold repair", such as bonded repair, is an obvious solution, due to a short down-time and non-intrusively process. However, currently no standards or rules exist for this kind of repair and engineering faces problems as basic as strength qualification. To address the lack of knowledge on the strength assessment of bonded repair for primary structure, Bureau Veritas Marine & Offshore launched a Joint Industrial Project (JIP) named StrenghBond Offshore with oil companies, shipyards and suppliers. The main objectives of the JIP are to:

- Assess short term and fatigue strength of typical bonded repairs,
- Enrich knowledge of adhesive joints strength on typical offshore repairs cases,
- Enable a better evaluation of the margin between the actual strength of a repair and the design load,
- Validate the characterisation procedure for strength prediction of bonded assembly,
- Define a robust strength prediction method,
- Gather the collected experience in a industrially applicable guideline,
- Standardise qualification process for offshore composite bonded repairs.

The project intends to provide a design approach for bonded reinforcement that is design orientated, accurate and recognized by the offshore industry.

This paper presents the general approach developed in the JIP, based on large test campaign using model scale specimens and full scale mock-ups, and on numerical simulations among which cohesive zone modelling.

Keywords: FPSO, composite materials, cold repair, design, strength

INTRODUCTION

The maintenance of steel structure installed in harsh offshore environment is challenging for oil & gas companies and marine suppliers. Vessels and mobile offshore units can be maintained and repaired onshore in shipyards but fixed units such as platforms or FPSO (Floating Production, Storage and Offloading) do not come back in dry dock and shall be cared of at sea. As FPSO are mainly installed in tropical areas, with marine environment, high temperature and high humidity, corrosion is a permanent threat. Since the early 2000s, the number of FPSO units has been multiplied by 6, see Figure 1, and at the present time, half of the fleet is over than 10 years old. The classical "crop and renew" technique remains mainly for the moment the unique solution for owners. However, cropping a structure at sea implies a number of major issues for owners such as "hot work" and international regulations such as ISGOTT (ISGOTT - International Safety Guide for Oil Tankers and Terminals, June 2006), requests for obvious safety reasons to empty and inert not only the tank directly concerned by the hot work, but also the adjacent ones. For these reasons, crop and renew technique becomes not only an intrusive solution but also necessitates long down times for the units.

The need for bonded repairs has been clearly identified since decades (Jan R. Weitzenbock, 2005) and some solutions have been already recognized by the industry, such as SPS Technology[1] and recently Coldshield®. "Cold repair", such as bonded repair, is an obvious solution, due to a short down-time and non-intrusively process. However, currently existing standards or rules cannot fully support industrial applications and engineering faces problems as basic as strength qualification. To address this need, Bureau Veritas and a consortium of owners have launched Strength Bond Offshore Joint Industrial Project (JIP SBO) with the objective to develop enable qualification of composite bonded repairs for the offshore units but also for marine applications.

This paper presents the general approach developed in the JIP, based on large test campaign using model scale specimens and full scale mock-ups, and on numerical simulations among which cohesive zone modelling.

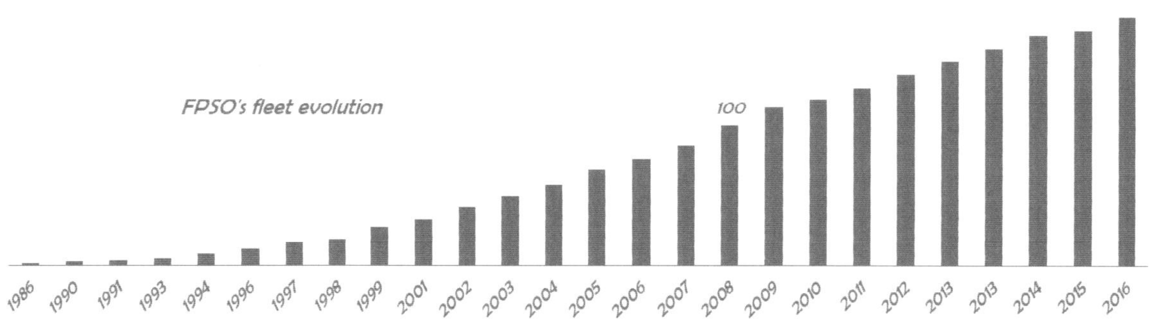

Figure 1: FPSO fleet evolution, (Worldwide Survey of Floating Production, Storage and Offloading Units, 2016)

STRENGTHBOND PROJECT

The JIP StrengthBond Offshore project initiates by Bureau Veritas Marine & Offshore (BV) has been launched in March 2019 for a duration of 3 years. Based on the long experience of BV in composite materials and adhesive joints (Bureau Veritas, 2015), the project aim to achieve the main following objectives:

[1] SPS Technology: only the welded-bonded version of the product is certified by Bureau Veritas at this date.

- Assess short term and fatigue strength of typical bonded repairs,
- Enrich knowledge of adhesive joints strength on typical offshore repairs cases,
- Enable a better evaluation of the margin between the actual strength of a repair and the design load,
- Validate the characterisation procedure for strength prediction of bonded assembly,
- Define a robust strength prediction method,
- Gather the collected experience in a industrially applicable guideline,
- Standardise qualification process for offshore composite bonded repairs.

The partners involved in the JIP StrengthBond Offshore are: Total, Petrobras, Naval Group, ColdPad, InfraCore and Siemens. A PhD student, in convention with Université Gustave Eiffel, is dedicated on tests and of numerical simulations along the project.

Project organisation

The project consists in 8 Work Packages, as presented in the Figure 2 below:

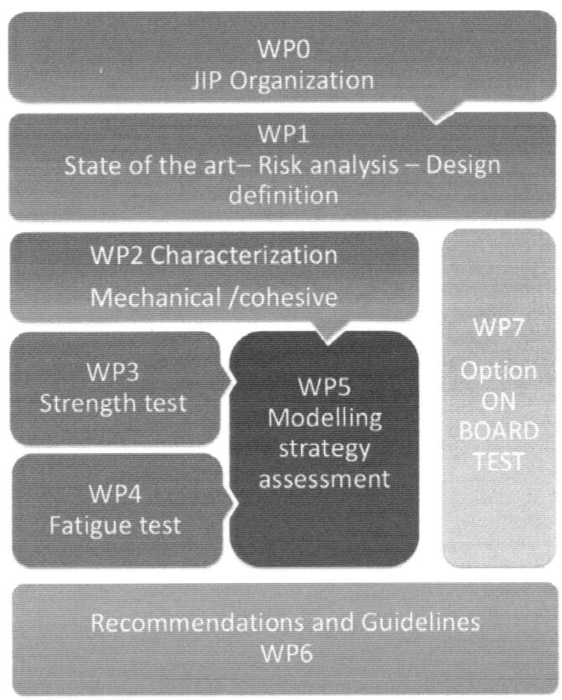

Figure 2: StrengthBond Offshore work package organization

The workflow of the project is articulated between characterisation (WP2), strength (WP3) and fatigue (WP4) tests, to be compared to a large modelling work investigating the failure prediction techniques (WP5).

The final document of the project will be an engineering recommendations and guidelines (WP6) focused on a robust methodology for the design assessment of patch repair for offshore and marine industries.

Patch repair context

Repairing a corroded deck of a ship shape unit imposes an important constraint on the design. The patch is laid in an area which is fully subject to hull girder stress. As it is ironically illustrated on Figure 3 (top), it is not possible to design the patch that ends out of the stressed area or a patch that wraps around some element to avoid peel stress.

This is a severe constraint as it imposes a high stress in the bond line edges. No simple solutions allow lowering the stress at the edges and the classical design techniques associated with large safety factors are not sufficient anymore. In those conditions, quality manufacturing and durability becomes mandatory and it becomes a difficult equation. Strength is the real driver for composite patch repairs design, however, many obstacles are inbound in the nature of composite patch repair for steel structure:

- Discontinuity of material, steel to composite,
- Non linearity of majority of adhesives,
- Singularity at the edge of the patch.

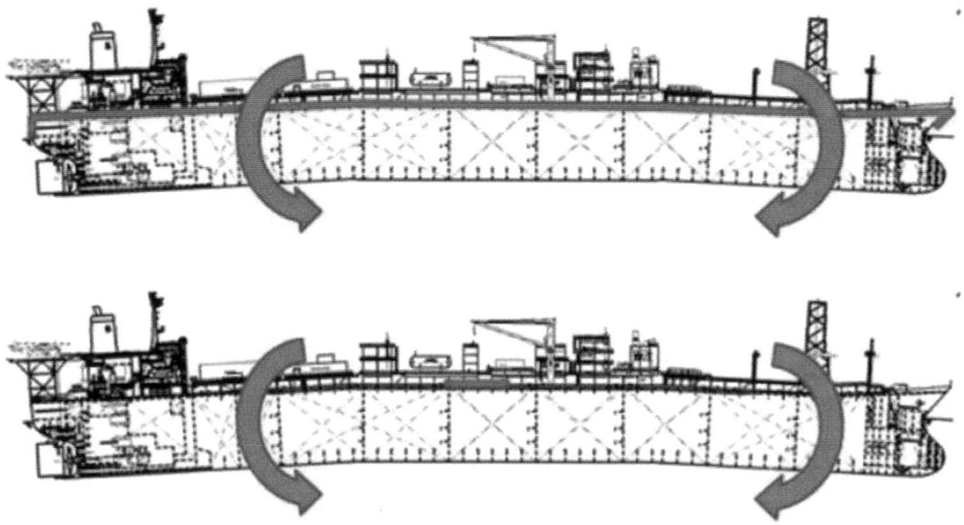

Figure 3: Ideal patch repair (top) vs realistic patch repair (bottom)

Strength prediction is difficult and stress based criterion are inaccurate in many cases as explained in (Mouton, 2018). Crack propagation techniques are too time consuming, and as a result laboratory tests remain the preferred option.

Patch repair design

The patch repair studied in the project is composed of several materials and layers as presented in Figure 4. For simulations inputs, all materials are to be characterized by tests.

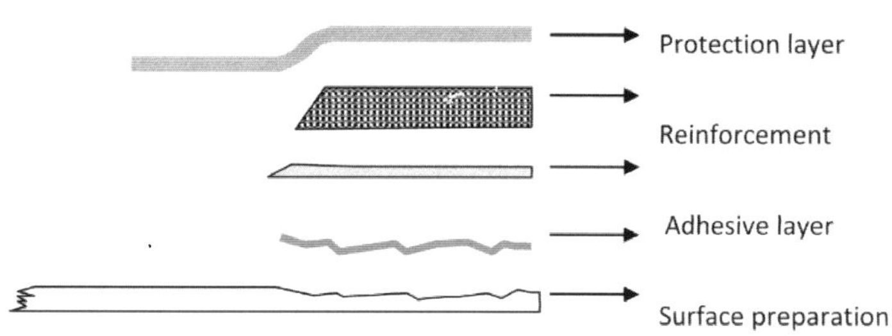

Figure 4: Patch repair system scheme

The following sections present characterisation tests that will be performed in WP2, mechanical/cohesive characterisation, and in WP3 strength tests, see Figure 5.

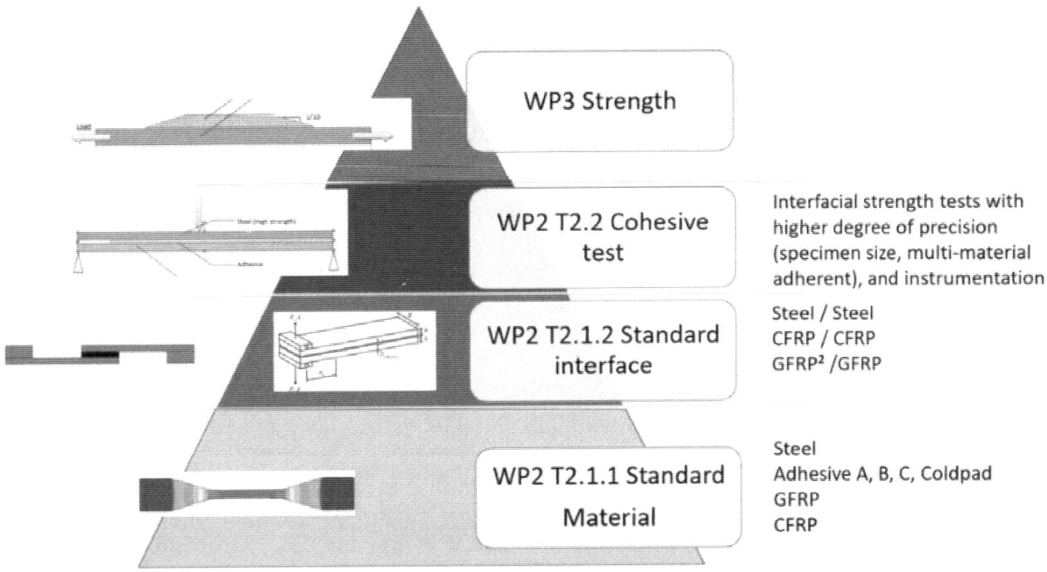

Figure 5: Tests pyramid

CHARACTERISATION TESTS

In the StrengthBond Offshore work package 2, adhesives and assembly should be fully characterized by performing:

- Standard materials tests: characterization of the normalized material properties such as Young modulus, Poisson coefficient, etc…,
- Standard and nonstandard interface tests: characterization of the cohesive laws of the assembly substrate adherent adhesive

Several materials have been selected however, at the end of the characterization, one set of adhesives/composite will be selected to be further assessed in the project and feed the modelling in the WP5.

Standard materials tests

The objective of those standard tests is to obtain mechanical properties of materials used for specimens for numerical simulation inputs. A highest accuracy of values used in numerical models will facilitate analyses and correlations with test results.

Steel characterisation

Specimen substrates are in steel. Tensile tests allow to determine following mechanical properties: Young modulus (E), Poisson coefficient (ν), yield (Ry) and ultimate (Rm) stresses.

Two steel substrates will be characterised using 5 coupons each, so 10 tests in total.

Resin bulk characterisation

Differential Scanning Calorimetry (DSC) allows to determine the glass transition temperature (Tg) of the resin itself. This value is important to know the acceptable temperature conditions for the optimal use of the resin.

Tensile tests are performed to obtain Young modulus (E), elastic limit (Ry), Poisson coefficient (ν) and ultimate strength (Rm) and strain (ε). The objective of tensile test is also to identify the behaviour type of the resin: elastic, plastic, hyper-elastic, brittle, etc.

TAST tests (Thick Adherent Shear Test) are required to evaluate the shear modulus (G), shear stress (τ) and shear strain (γ).

Tensile or flexural tests are to be performed to assess the creep behaviour, as well as diffusion test (D) and density measurement (ρ).

In total 4 different resins, epoxy and vinylester, will be tested in accordance with recognized standards to obtain those properties leading to 100 standards tests.

Composites characterisation

The patch will be composed of two kind of fibres, glass and carbon. The main function of the glass layer is to insulate the carbon to the steel substrate and to protect against galvanic corrosion. The characterisation of the composite laminates consists in tensile and compressive tests in 0° and 90° fibre direction. Mechanical properties in both direction, longitudinal and transverse, such as Young modulus (E_{1t}, E_{2t}, E_{1c}, E_{2c}), shear modulus (G_{12}), Poisson coefficient (ν_{12}, ν_{21}) and maximal strength (X_{1t}, X_{2t}, X_{1c}, X_{2c}) are measured as well as shear stress (τ).

In addition to the DSC tests for the Tg determination, some physico-chemical tests must be realised to measure the reinforcement content (Vf) and the density (ρ) of laminates.

In total, 68 standards tests have been identified and should be carried out by the lab for the composite characterisations.

The Figure 6 shows a schematic representation of the steel substrate with the composite patch in glass fibre (GFRP) and carbon fibre (CFRP) and the mechanical properties obtained by tests for each material. Table 1 summarises the number of tests to be performed for materials characterisation.

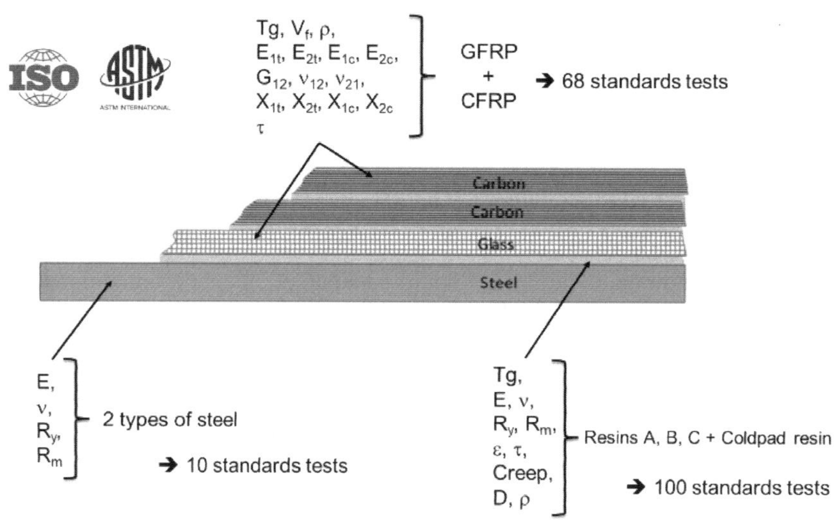

Figure 6: Summary of standard material characterisation tests

Table 1: Materials standard tests

Materials	Number of types	Number of test
Steel	2 grades	10
Resin	4 (epoxy and vinylester)	100
Composite	2 (glass and carbon)	68
Total		178

Standard interface tests

Previous tests are not sufficient to simulate the debonding of the patch at the steel substrate and the delamination of laminates. For using cohesive elements in numerical models, some additional information are to be determined by specific tests for the interface characterisations.

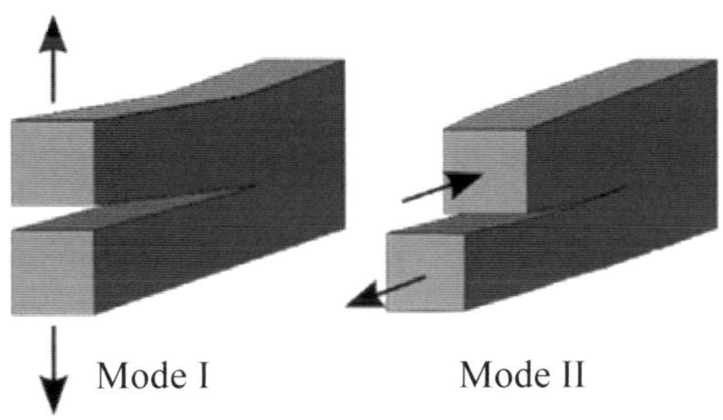

Figure 7: Mode I and Mode II

The objectives of those tests are to obtain the fracture strength of adhesive in bonded metal joint and the interlaminar fracture toughness for mode I, mode II and mixed mode, see Figure 7.

Interface characterisations

The Figure 8 shows a schematic representation of the steel substrate with the composite patch in glass fibre (GFRP) and carbon fibre (CFRP) and characteristic values obtained from standard interface tests.

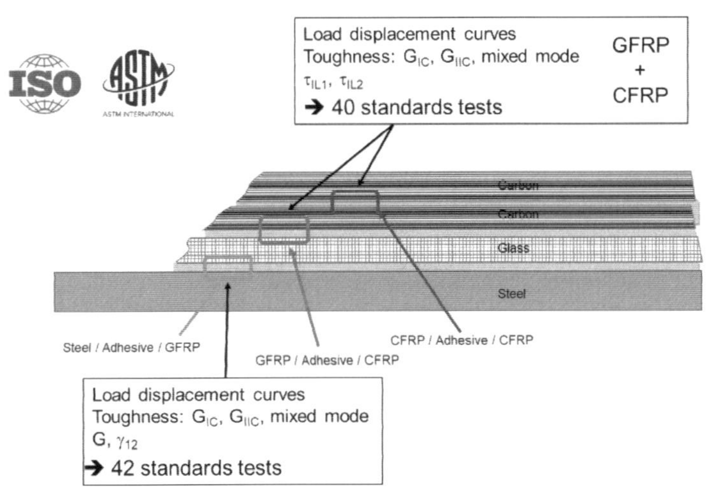

Figure 8: Summary of standard interface characterisation tests

As the thickness of the resin, considered as the adhesive, is thin, three kind of interfaces have been selected:

- Steel/Glass fibre interface
- Glass fibre/Carbon fibre interface
- Carbon fibre/Carbon fibre interface

Following selected standard tests are to be performed and detailed hereafter:

- Double Cantilever Beam (DCB),
- End Notch Flexure (ENF),
- Mixed Mode Bending (MMB).

DCB tests – Mode I

The propagation of the crack in mode I is carried out by an opening of the specimen in an interlaminar plane. The DCB test consists of gluing two test pieces together with a non-adhesive shim in way of the load application. The spacing of the load application point during the test then leads to a propagation of the crack according to a mode I. Figure 9 schematizes the DCB test to be carried out on a test piece in order to obtain the fracture toughness G_{Ic}, the adhesive critical energy release rate in mode I.

The tests are standardized, they can be performed according to the ASTM D3433 standard.

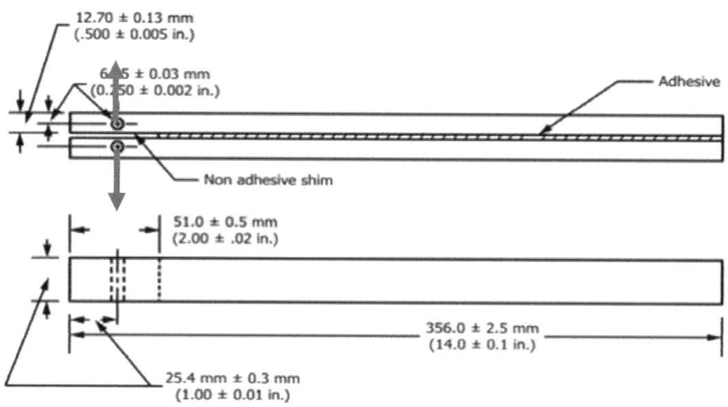

Figure 9: DCB test, ASTM D3433

ENF tests – Mode II

ENF test consists in applying a 3-point bending test to a test piece notched at its end. The propagation of the crack is carried out under the effect of interlaminar shear stress. This type of test gives an estimation of the shear toughness G_{IIc}.in mode II.

Figure 10 schematizes the ENF test as required with standard ASTM D7905.

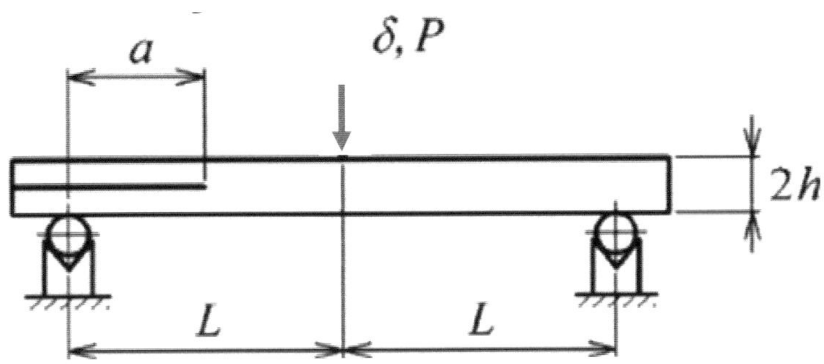

Figure 10: ENF test, ASTM D7905

MMB tests

The MMB test allows to perform a crack propagation test using a mixed mode I and mode II. This test is based on the DCB and ENF tests, where the tested toughness ratio G_{Ic}/G_{IIc} is a function of the geometric parameter c and c_g, in accordance with ASTM D6671, see Figure 11.

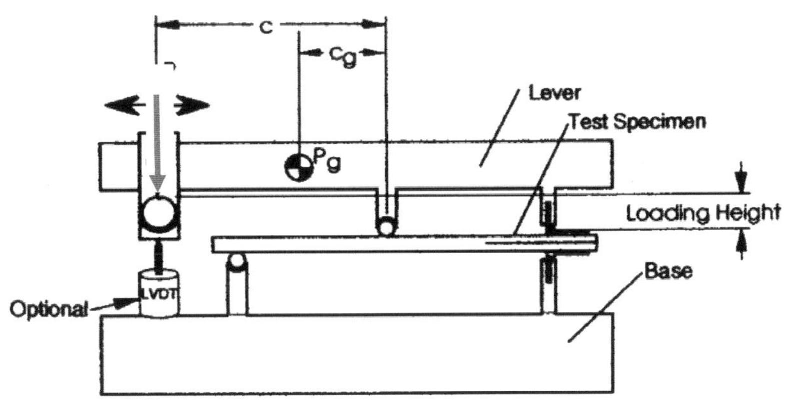

Figure 11: MMB test, ASTM D6671

Table 2 summarises the number of standard tests to be performed for interface characterisation.

Table 2: Interface standard tests

Interface	Number of test
Steel/adhesive/GFRP	42
GFRP /adhesive/CFRP	20
CFRP /adhesive/GFRP	20
Total	82

Non-standard interface tests

Although previous tests aims at obtaining shear toughness in mode I, mode II and mixed mode, no standards define a precise manner of extracting the cohesive law parameters. A state of art analysis have been performed in WP1 in order to set the law parameters and to ensure a proper model calibration for further simulations.

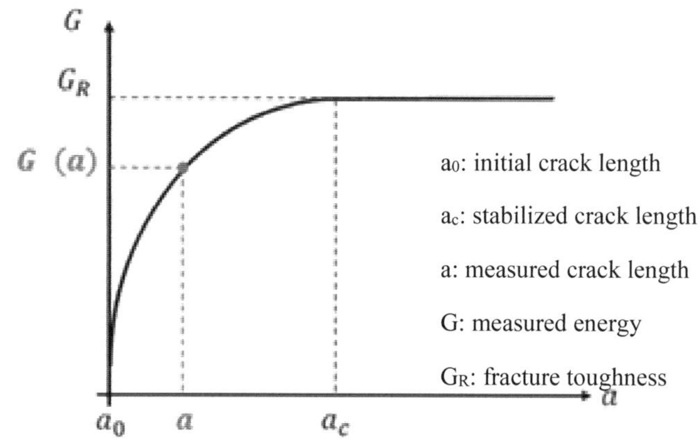

Figure 12: Typical energy measurement curve

As the objective is to characterize the debonding for numerical simulations, the correct adhesive fracture energy is to be quantified by test. However, as there is no direct standard measurement techniques, energies in the system are to be evaluated by measuring load, displacements and crack length. As illustrated in Figure 12 by a typical energy measurement curve, the stabilized energy value is obtained after a certain crack length which can be longer than the minimum required specimen length. The interface toughness G_R is the one from the plateau. In order to reach the stabilized energy toughness value, it is proposed to increase the sample length to ensure a good toughness stabilization and compare results to standard test values.

The sample width can be also a problem due to the direct relation within the crack process zone shape and the stress/strain state (Begue, 2019). To address these issues, new samples have been defined in addition of standard interface test specimens. The goal being to:

- Study the plan stress/plan strain ratio effect on bonded patch,
- Ensure that the standardly obtained G_R are not dependent to the sample size.

Also, the bonded patch is composed of multiple layer of GFRP and CFRP, of woven or unidirectional plies (set in different direction), the common approach is to obtain the critical energy release rate of each type, in mode I, II and mixed mode, to ensure a correct modelization, however, it leads to really large test campaign. To limit the number of test, a dedicated specimen has been developed to consider directly and representatively the material assembly, see Figure 13. This design will allow to perform DCB test, for mode I, ENF test, for mode II, and MMB test, for mixed mode, with the actual complex interface design. Test results shall confirm the possibility to assess in one test the lowest interfacial toughness with the potential to enable significant reduction of specimen number.

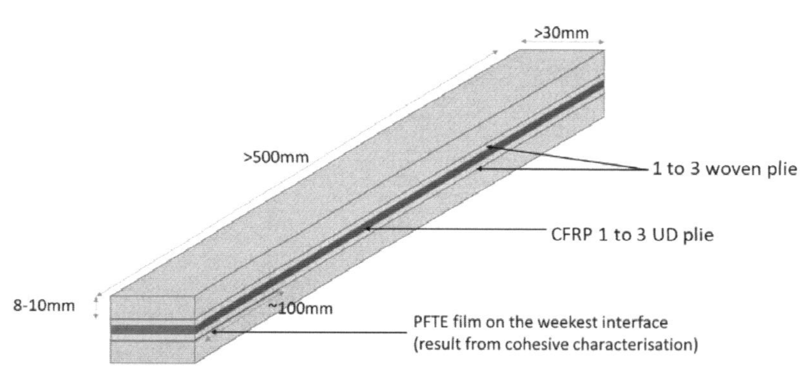

Figure 13: Equivalent interface sample geometry

Non-standard specimens will be monitored with Rayleigh optical fibre, in and out of the composite substrate. Such optical fibres allow a continuous strain measurement along samples, and are able to detect crack initiation or follow crack propagation. Moreover, a DIC system (Digital Image Correlation), a LVDT (Linear Variable Differential Transformer) will be installed for local surface strain and displacement measurements.

STRENGTH TESTS

The objectives of the strength tests in the work package 3 of the JIP StrengthBond Offshore project, are twofold:

- Assess the strength of representative offshore units repairs and explore the influence of the design and installation parameters;
- Constitute a base of experimental data with different loadings and designs in order to provide the necessary data to calibrate scantling methods (WP5) and mostly to assess the reliability and robustness of the calibrated method on several different configurations on design and loading.

In the scope of the first objective, test specimens are to be as representative as possible of the offshore units repair design and most challenging loads, i.e. hull girder strength. Single strap tests, as shown on Figure 14, allows to represent the actual hull girder loading strength of offshore patch repairs. Single strap specimens enable to test representative patch extremities but do not allow to transit large loads through the bond line as there is no rupture of the steel load path. On the contrary set-ups such as double lap shear or double lap joint, see Figure 15, present a discontinuous steel load path, so 100% of the load transfers through the bond line. Those types of set-up do not allow to represent pad extremities as per the single strap as the maximum stresses in the bondline is expected at the centre. Being completely different loading mode it constitutes a stringent evaluation of the robustness of the strength prediction method.

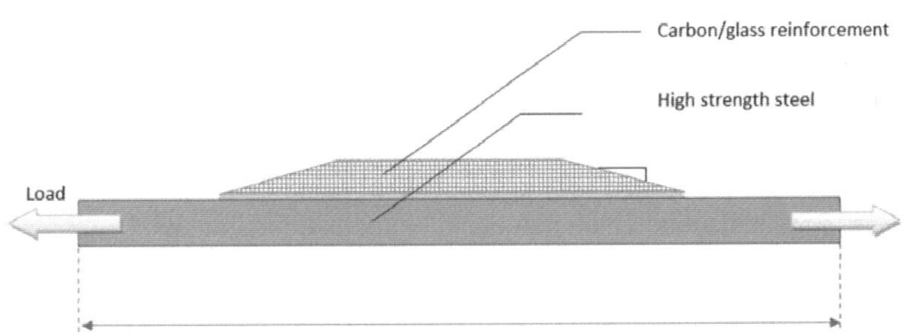

Figure 14: Single lap specimen

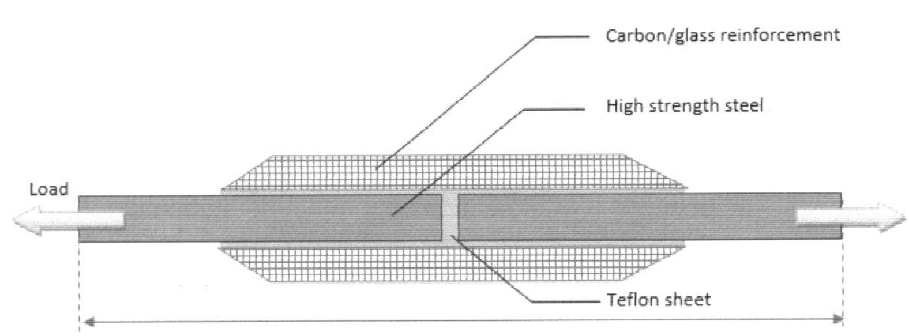

Figure 15: Double lap specimen

For all tests data are to be recorded and documented on failure load, failure mode, crack propagation, standard deviation, load-displacement curve, strain curves… and used for the numerical simulation correlations.

NUMERICAL APPROACH

The final goal of all these tests is to be able to simulate with the highest accuracy the mechanical behaviour of a bonded composite patch repair and the debonding. The work package 5 focuses on three objectives:

- Build up, assess and compare failure prediction methods for bonded repairs,
- Assist in understanding the test results by producing numerical twins of the test in which data such as interfacial stresses can be precisely read,
- Define and apply methodology for strength assessment of repair and efficiency estimation.

The achievement of these objectives is based on the material and interface characterisation tests performed in WP2 and on the results of strength tests in WP3.

Cohesive elements

In numerical simulations, the cohesive zone modelling consists of the definition of a zone in the adhesive, at the interface or the whole adhesive (continuum approach) in which a damage law is implemented. The damage law is characterized by at least 3 parameters conferring to the interface layer:

- A stiffness module,
- A critical strain/stress,
- A total energy release rate until complete failure (critical energy release rate).

Such techniques have been assessed with good results in previous literature (Shahrokh, 2012), (Baker, 1988), and in research project Co-Patch (Co-patch, 2010). However, no paper was found covering the full test techniques for calibration of the parameters and validation on actual bonded repair with different configuration/designs. This is considered as a major objective for the project, and a large step forward in the industry, to assess the full process of validating a bonded structure with cohesive element purposely characterized by tests. The influence of the mesh size and the type of cohesive laws, bilinear as Figure 16, or polynomial law, will be assessed and analysed as well as the calibration of the toughness determined by characterisation tests. Strength test simulations will be compared with monitored specimens and will be tuned to represent the reference set-ups. The failure strategy will be based on these results and implemented in a full scale model.

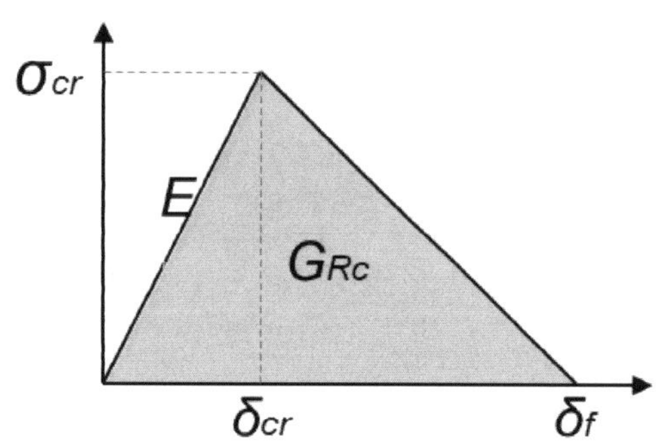

Figure 16: Bilinear model of the traction separation law in cohesive zone modelling

Case study

Using findings of the previous steps, a representative case study will be addressed to confirm the strength prediction methodology developed with the test and the adapted model strategy. This case study will be chosen in accordance with the owners in the JIP as representative of the needs within the scope of the JIP StrengthBond Offshore.

The analysis is to be carried out based on the typical rules used in class approval or design stage. The case study should assess the adequacy between the solution develop during the JIP with the industrial application. The efficiency of the design shall be assessed in order to validate the resistance based on rules and to establish corresponding safety factors.

PRELIMINARY RESULTS

Experimental ENF results

First ENF tests have been performed by University Gustave Eiffel on a highly instrumented specimen combining continuous optical fibre and Digital Image Correlation (DIC), see Figure 17.

A 150µm-diameter optical fibre with polyimide coating has been bonded on upper surface of the specimen and inside the bond line in the centre of the specimen. Coupled with the Optical Distributed Sensor Interrogator Luna ODiSI-B, strains can be measured along a fibre, up to 10 m long with maximal spatial resolution of 1.2 mm. For testing, frequency sets at 12 Hz, recording the strain during all the test, and all along the bonded optical fibre.

Two cameras have been installed in order to monitor a size corresponding to 6cm x 2cm around the initial crack location. For testing, a mesh of 32 pixels * 32 pixels allowing to have a measurement resolution of 0.5 * 0.5 mm has been used. The local displacement measurement has a resolution of 5µm at a recording frequency of 5 Hz.

Figure 17: Instrumented ENF specimen

Results obtained with these tests have been deeply analyzed by Université Gustave Eiffel and will be presented in another conference paper showing the advantage of using continuous optical fibre for the analysis of ENF tests (Sourisseau Q., 2021). Indeed, the optical fibre being located inside the specimen enable to determine very precisely the initial crack front and the strain state of the bondline. The external fibre can be used to determine the crack front propagation during the test. Those accurate measurements combining with different theory: Compliance Calibration Method (CCM) (Davies, 2001), ASTM D7905 CCM or Linear Elastic Fracture Mechanics (LEFM), allow determining the critical toughness G_{IIc} of bonding joint (Table 3) to be used for cohesive zone element in numerical simulations.

Table 3: Determined G_{IIc}

Method	G_{IIc} (kJ/m²)
CCM	0.87±0.02
ASTM D7905 CCM	0.71
LEFM	0.8

Numerical ENF results

A Finite Element Model has been carried out by Siemens to represent the ENF test (Figure 18) in order to correlate numerical model with testing specimen and confirm determined toughness value.

The three-dimensional model is built with monolayer brick volume for substrate parts, and both the adhesive layer and the adhesive/substrate interface are modelled by cohesive elements. A previous study showed that for these materials and these specimen ratios, there were no interest in modelling adhesive layer separately via other brick volume. Due to the symmetry plane, only a half of the structure is modelled. Mesh is refined lengthwise at end of pre-crack and in contact zones and mesh density is laterally more important at free edges to catch shearing. Thus 2510 bricks and 480 cohesive elements were used.

Node to facet contact conditions were added between substrate lips of the pre-crack. Supports and loading tool were modelled as rigid surfaces cylinders, they are linked to the structure by contact conditions too.

From experimental tests, a stress/strain relationship was identified, and then critical energy rate was deduced. From the shape of the curve, it comes a polynomial law for which a couple of parameters were obtained: stiffness and exponent. Parameters were validated using an elementary model before to be used in this simulation.

Load is applied by controlling cylinder tool displacement, case is solved by a non-linear quasi-static solution included in Simcenter Samcef 2020.2.

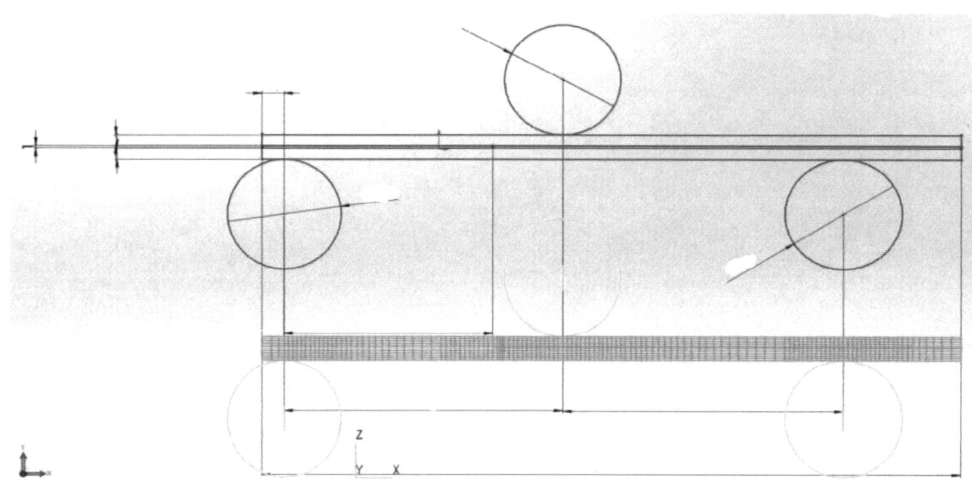

Figure 18: ENF test scheme (top) and model (bottom)

Analytical (dashed curve), numerical (blue and brown curves) and experimental (black curve) results have been compared as plotted in Figure 19. Analytical solution including transverse shear deformation contribution was used as a validating tool (plotted in Figure 19 for Gc value 0.87 N/mm²). A good correlation was obtained for elastic slope (partly determined by cohesive elements stiffness) between experimental and numerical results and in a less extent with analytical model. A slight discrepancy occurs regarding delamination initiation (maximum load value) between numerical (for Gc value 1.06 N/mm², brown curve) and experimental values. From a deeper analysis of experimental results, a new critical energy value was identified and then applied in numerical simulation (for Gc value 0.87 N/mm², blue curve). A better correlation for initiation and propagation phases is then stated. As a comparison, another simulation was performed using an exponential law (and a Gc value 0.87 N/mm²). The discrepancy between experimental and numerical results confirm the right identification of polynomial law parameters as well as the importance of a correct analysis of experimental data.

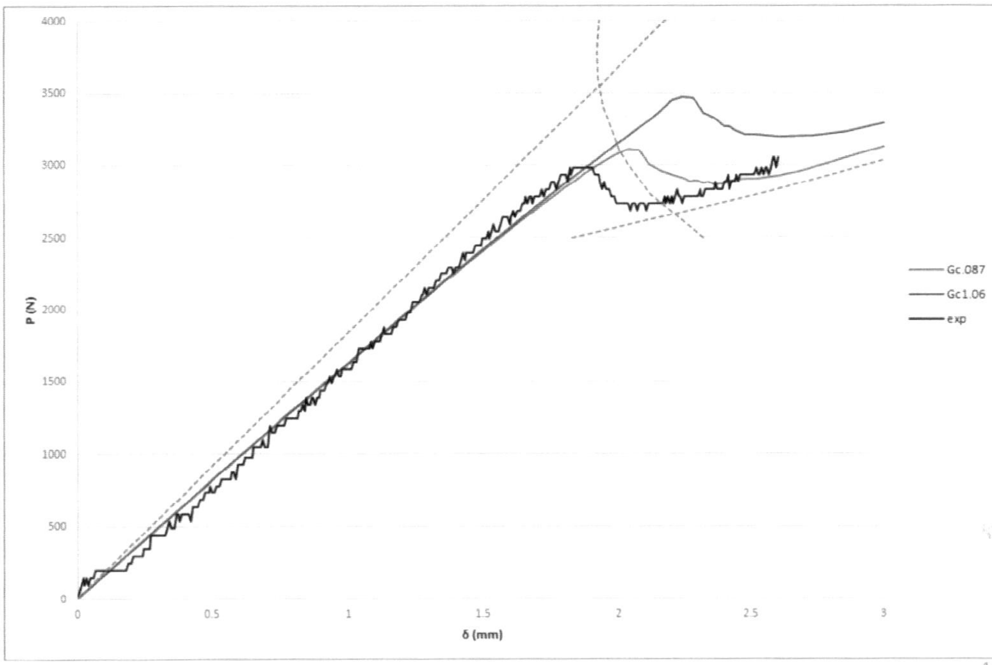

Figure 19: Analytical, numerical and experimental results comparison

A sensitivity study on cohesive law parameters will be performed in order to evaluate their influence on results.

CONCLUSIONS

As detailed in this paper, a large campaign of tests will be performed within the JIP StrengthBond Offshore project. More than 300 specimens will be tested for the material characterisation and for the interface characterisation. As standards seem insufficient to provide information for the determination of cohesive law, new specimens have been defined to address this issue. In addition, a dedicated specimens, representative of the patch assembly has been developed for the characterisation.

Once the mechanical properties are known, strength prediction solutions are to be defined and implemented in numerical simulations. The comparison of strength tests and model results will confirm the robustness of the approach and preliminary results seem to confirm this.

The failure strategy developed within the project suggests an efficient and accurate methodology for the design assessment of bonded repair for marine and offshore applications.

This work will be the basis for further assessment of ageing and durability of composite patch repair and pave the way to qualification of bonded repair solutions.

ACKNOWLEDGEMENTS

Authors wish to acknowledge the technical and financial support of JIP StrengthBond Offshore partners: Total, Petrobras, Naval Group, ColdPad, InfraCore and Siemens.

REFERENCES

ASTM D3433, Standard Test Method for Fracture Strength in Cleavage of Adhesives in Bonded Metal Joints

ASTM D6671, Standard Test Method for Mixed Mode I-Mode II Interlaminar Fracture Toughness of Unidirectional Fiber Reinforced Polymer Matrix Composites

ASTM D7905, Standard Test Method for Determination of the Mode II Interlaminar Fracture Toughness of Unidirectional Fiber-Reinforced Polymer Matrix Composites

Baker, A. (1988). *Bonded Repair of Aircraft Structures.* Martinus Nijhoff Publishers.

Begue, J.-F. (2019). *Fatigue & Fracture Mechanics.*

Bureau Veritas. (2015). NI613 Adhesive Joints and Patch Repair.

Co-patch. (2010). Retrieved from http://www.co-patch.com/.

Davies, P. (2001). Fracture Mechanics Testing Methods for Polymers, Adhesives and Composites. *Adhesives and Composites, Chapter: Introduction to delamination fracture of continuous fibre composites elsevier Science. 28*, 271-275.

ISGOTT - International Safety Guide for Oil Tankers and Terminals. (June 2006). 5th Edition.

Jan R. Weitzenbock, D. M. (2005). *Bondship, Project Guidelines.*

Mouton, L. (2018). Enhance reliability of structural bonding: an advance solution of repair for corrosion onboard offshore units. *37th International Conference on Ocean, Offshore and Arctic Engineering, OMAE2018.* Madrid.

Shahrokh, z. (2012). Near-treshold Fatigue of Adhesive Joints: Effect of Mode Ratio, Bond Strength and Bondline Thickness.

Sourisseau Q., L. E. (2021). Use Of Distributed Optical Fibre To Monitor The Crack Propagation Of An Adhesively Bonded Joint During An ENF Test. *10th International Conference on FRP Composites in Civil Engineering, CICE.* Istanbul.

Worldwide Survey of Floating Production, Storage and Offloading Units. (2016). *Offshore Magazine.*

APPLICATION OF FAST SYNCHROTRON X-RAY IMAGING IN VELOCIMETRY OF CAVITATING FLOWS

MINGMING GE
KEVIN T. CROFTON DEPARTMENT OF AEROSPACE AND OCEAN ENGINEERING, VIRGINIA TECH, BLACKSBURG, VA 24060, USA

GUANGJIAN ZHANG
RESEARCH CENTER OF FLUID MACHINERY ENGINEERING AND TECHNOLOGY, JIANGSU UNIVERSITY, ZHENJIANG 212013, CHINA

NAVID NEMATIKOURABBASLOO

KEVIN T. CROFTON DEPARTMENT OF AEROSPACE AND OCEAN ENGINEERING, VIRGINIA TECH, BLACKSBURG, VA 24060, USA

KAMEL FEZZAA
ADVANCED PHOTON SOURCE, ARGONNE NATIONAL LABORATORY, 9700 S. CASS AVENUE, LEMONT, IL 60439, USA

OLIVIER COUTIER-DELGOSHA
KEVIN T. CROFTON DEPARTMENT OF AEROSPACE AND OCEAN ENGINEERING, VIRGINIA TECH, BLACKSBURG, VA 24060, USA

Proceedings of the 26th Offshore Symposium, April 6&7th 2021, Houston, Texas
Texas Section of the Society of Naval Architects and Marine Engineers

ABSTRACT

Hydrodynamic cavitation is a complex two-phase flow phenomenon involving mass and heat transfer between liquid and vapor. It occurs in many widely-used hydraulic machines, such as pumps and marine propellers, and often leads to undesired effects like material erosion, noise, and vibration. To control these detrimental effects, the visualization of two-phase flow morphology inside the opaque cavity is a crucial point to improve the physical and numerical models of cavitation. The major challenge in experimental measurements of cavitating flow fields is the fact that multiple scattering and a direct reflection of visible light from phase boundaries make the flow optically opaque. In recent years, unlike traditional local measurements using various probes, the development of the third-generation synchrotron radiation sources promotes the application of X-ray phase-contrast imaging, which enables clear visualization of boundaries between phases with different refractive indices. In this study, the partial cavity is formed in a convergent-divergent (Venturi) channel with a small contraction ratio where the relatively stable cavitation regime can be sustained in a wide range of cavitation numbers. The experiment performed at Advanced Photon Source (APS) of Argonne uses the short high-flux X-ray pulses emitted from synchrotron sources to capture fast dynamic events and minimize motion blur. The internal two-phase structures and bubble development dynamics inside the quasi-stable sheet cavitation can be identified. Aside from the detailed illustration of two-phase morphology, X-ray phase-contrast images were also used to perform velocimetry by tracking either seeded particles or phase interfaces inside the opaque regions. Through appropriate postprocessing to the recorded X-ray images of cavitation, the time-resolved velocity and void fraction fields are obtained simultaneously. These unprecedented experimental data will be further explored in understanding fluid mechanics underneath the cavity, estimating slip velocity between fluid-vapor interactions, and reconstructing pressure fields for compressible flows.

Keywords: Cavitation, Two-phase flow, X-ray imaging based velocimetry

INTRODUCTION

Cavitation is a phenomenon in which rapid changes of pressure in a liquid lead to the formation of small vapor-filled cavities or bubbles. The physical process of cavitation can be compared with boiling in which the evaporation occurs when the local temperature of the liquid reaches saturation temperature. While cavitating vapor appears even at room temperature when the local pressure falls sufficiently far below the saturated vapor pressure. In most engineering cases the issues with cavitation are related to hydrodynamic cavitation, with different forms including single bubble cavitation, streak cavitation, tip vortex cavitation, partial and supercavitation, etc. The large-scale vapor cloud shedding under certain conditions is often responsible for undesired effects like performance degradation, erosion damage, and system vibration.

To understand the internal structures of cavities and control these detrimental effects, experimental studies with various measurement techniques have been developed to investigate this complex phenomenon. Due to the existence of non-transparent liquid/vapor mixtures, a visual observation by a fast speed camera (high-speed photography) is the most straightforward and widely-used method to capture the temporal evolution of cavitation structures, thereby providing insight into the underlying physics. Dular et al. [1] observed the scale effect on the vapor shedding process in a small size Venturi tube based on high-speed photography images and spectral analysis of image grey level. Although high-speed photography is an effective way to visualize the global behavior of cavitation [2], it is difficult to measure the structure and dynamics inside the cavity sheet due to its opacity.

Stutz & Reboud [3-5] used a double optical probe to measure the time-averaged void fraction and vapor-phase velocity inside the cavity generated in a two-dimensional Venturi-type section. In this technique, the local velocity was estimated by the time interval of a bubble passing two probe tips successively, and the local void fraction was defined as the ratio of the cumulated time of vapor phase at the tip of the probe to a given time of observation. Despite a relatively large measurement uncertainty of about 15%, their work gave a preliminary description of the two-phase flow structure inside a sheet cavity and confirmed the presence of a reverse flow along the solid surface. Unlike pointwise measurements using intrusive probes, Particle image velocimetry (PIV) enables the rapid collection of instantaneous velocity vectors covering the entire flow field. And the issues related to the cavitation opacity are partially eliminated thanks to the combined use of laser-induced fluorescent (LIF) particles and optical filters. However, when intense vapor bubbles accumulating between the laser sheet and the camera, the light path starting from the fluorescent particles will be deviated severely or blocked completely. Consequently, most PIV-LIF measurements are applied in the regions outside the cavity sheet [6, 7] or turbulent shear layer with a low void fraction (less than 5%) [8, 9].

Another limitation for the light refraction measurements, for example using visible light or laser in PIV methods, is their low penetrability. As a comparison, X-rays owning a much shorter wavelength than visible light can penetrate most optically opaque media with weak interactions. This distinct advantage makes X-ray radiography a powerful method for studying opaque multiphase flows. For cavitating flows, the void ratio field which is important to understand the two-phase structures can be measured based on X-ray attenuation [10-12]. A standard 2D X-ray densitometry system can only provide a projection of the sample's density distribution in the direction of the X-ray beam. And velocimetry is difficult to be performed since the seeded particles can't be detected with this relatively low-level energy. To perform velocimetry by tracking either seeded particles or phase interfaces, the third-generation synchrotron radiation sources such as the advanced photon source (APS) at Argonne National Lab can be applied. With a high-energy and coherent X-ray beam, ultra-fast X-ray phase-contrast imaging has been applied to cavitating flows on a 2D Venturi-type section in our group since 2009 [13-15]. This technique makes it possible to obtain simultaneously: (I) high-resolution and time-resolved void fraction field, and (II) instantaneous velocity field of two phases in a single experimental system. With newly designed module-assemble test section and higher resolution of improved X-ray phase-contrast imaging, the consecutive evolution of cavity sheet inner structure is the first time measured with a much shorter time interval.

In this experiment, the cavitation is generated in a 3D-printed venturi-type flow channel, with a heating and cooling functioned loop to precisely adjust the water temperature. The flow rate is controlled by a frequency variation device connected to the circulation pump. Cavitation conditions can be set properly by regulating the pressure on the free surface of the tank utilizing a vacuum pump. Details about the Hydraulic test rig can be referred to in the works [16]. With the raw x-ray generated images, a method of extracting liquid tracer particles from vapor phase structures is proposed. The statistical results based on the time-resolved void fraction field and velocity field are presented and analyzed. The image processing and data analysis results show that the methods provided here can obtain a clear observation and quantitative analysis of the cavitation dynamics and shedding behavior.

METHODOLOGY

The experiments were performed using the third-generation synchrotron radiation at the Advanced Photon Source (APS) in the USA where a high-energy and spatially coherent X-ray source is available. As illustrated in Figure 1, the X-ray source is aligned with the test section on one side and the X-ray detector (a scintillator) on the other side converting the X-ray beam into visible light which is then recorded by the high-speed CCD camera. The acquisition frequency of the camera is set to 12070 fps, which enables a spatial resolution of 704×688 pixels with a pixel size of 2 μm. To obtain appropriate pairs of images for PIV analysis, the time interval between two consecutive images shall be reduced using the camera frame transfer method. Thus, a mechanical rotating chopper is used as the fast shutter to limit the time interval as small as 3.68μs. With this time interval and a 15m/s flow speed, the moving distance of a particle is 55μm or 28 pixels on the image. The other slow shutter, operating at a frequency of 1 Hz with an opening time of 24 ms, is equipped to protect the test section and the detector by reducing the heat load on them. The test section is a small size Venturi-type flow channel that was installed in a portable closed hydraulic loop. With a convergent angle of 18° at the inlet and a divergent angle of 8° at the outlet, cavitation is generated. Flowing through the test section, the cavitated multiphase flow was generated from the Venturi throat. Together with cavitation bubbles, silver-coated hollow glass spheres with nominal mean diameters of 10 and 17μm were injected into the flow as tracers of the liquid phase.

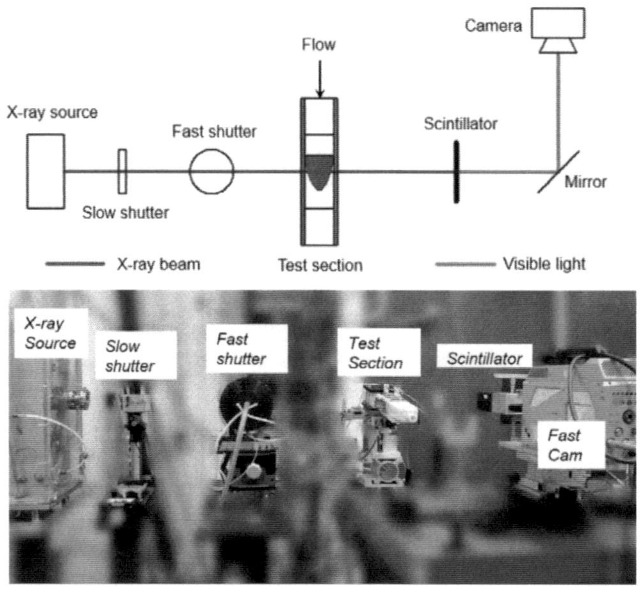

Figure 1: Lightpath of the X-ray imaging system at the APS (upper: diagram, lower: image)

A typical x-ray converted image is shown in Figure 2. Due to the absorption contrast and phase contrast mechanisms, the vapor bubbles and seeded particles can be distinguished from the liquid background. The absorption contrast resulting from the difference in the attenuation of X-ray energy after penetrating vapor, liquid, or glass particles with different absorption coefficients. Due to this absorption contrast, the internal intensity of objects (bubbles and particles) is higher than the surrounding liquid. As can be seen in the red circle, the sharp contrast at the periphery of a bubble is because of the Fresnel diffraction and the interference between neighboring points of the wavefront at a certain distance from the sample, which is called the phase contrast effect. To make phase effects detectable, an appropriate propagation distance between the object and the detector is required and the X-ray beam must be (at least partially) spatially coherent. For the image processing, the instrument function is not taken into account.

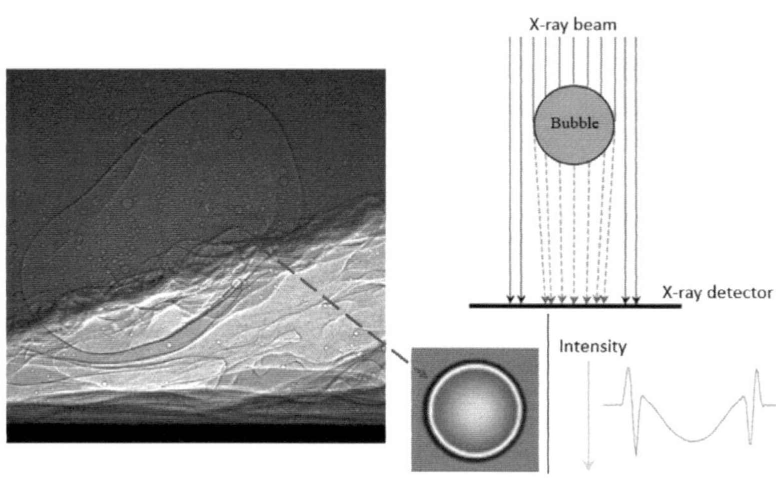

Figure 2: Detectable edges of bubbles based on phase contrast mechanism

RESULTS

As shown in Figure 3, the complete image for a sheet cavity is reconstructed by successive but not simultaneous acquisitions at different positions. That is because the effective illuminating cross-section of the X-ray beam is limited to 1.7mm×1.3mm, which is smaller than the cavity length around 20mm. To analyze the dynamics of each phase, image processing is required to extract the tracer particles from the liquid background and vapor structures so that velocimetry algorithms can be applied to each phase separately. In fact, from each X-ray radio-graph, three new images need to be created: (1) an image of particles for measurements of the liquid phase velocities, (2) an image of bubbles for vapor phase velocity measurements, and (3) an image of vapor structures with neither interfaces nor particles used for vapor volume fraction measurements [15].

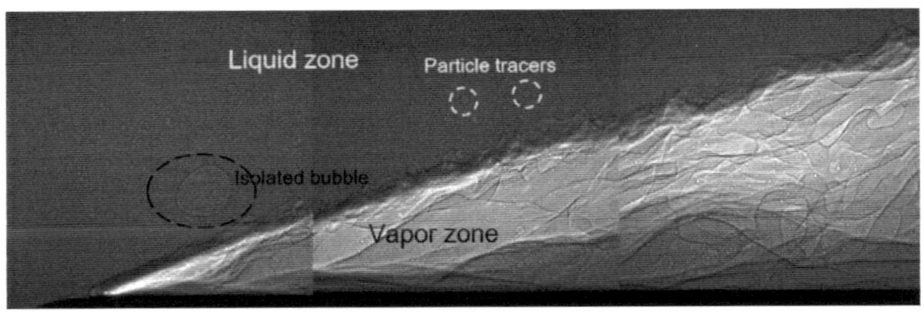

Figure 3: A cavity sheet reconstructed from the first 3 positions

Different image processing algorithms have been developed for the identification of tracers in our group [16-18]. Since a general or local threshold of the gray level was found inadequate to distinguish the objects, the edges generated from phase-contrast effects need to be detected. In the first step, edge detection using the Canny method was performed directly on the raw image. Followed by morphological operation, noise delamination, and watershed of coalesced objects, 652 particles can be detected as in Figure 4. However, the particles inside the vapor zone are not detected properly.

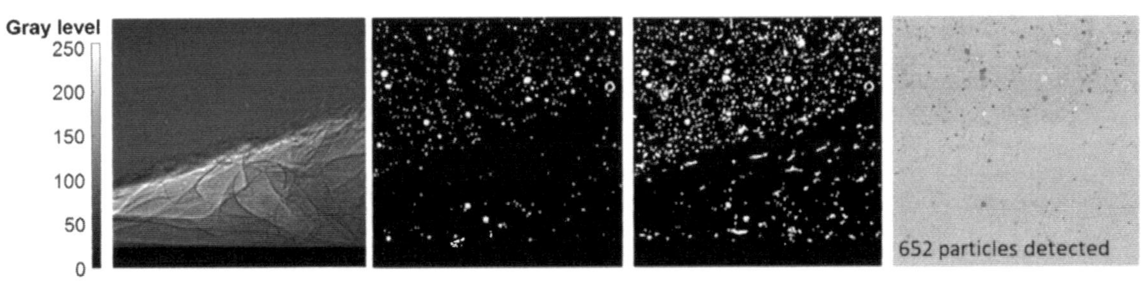

Figure 4: Edge detection on X-ray image without contrast enhancement

Therefore, before the interface detection process, enhancing the contrast of images would make this recognition more efficient. In this study, the FFT-based Gaussian filtering and wavelet filtering are used and results are combined. With Gaussian filtering, the high-pass image with interface information can be enhanced. The edge detection and morphology operation were applied on the contrast-enhanced image and 682 particles are detected as shown in Figure 5.

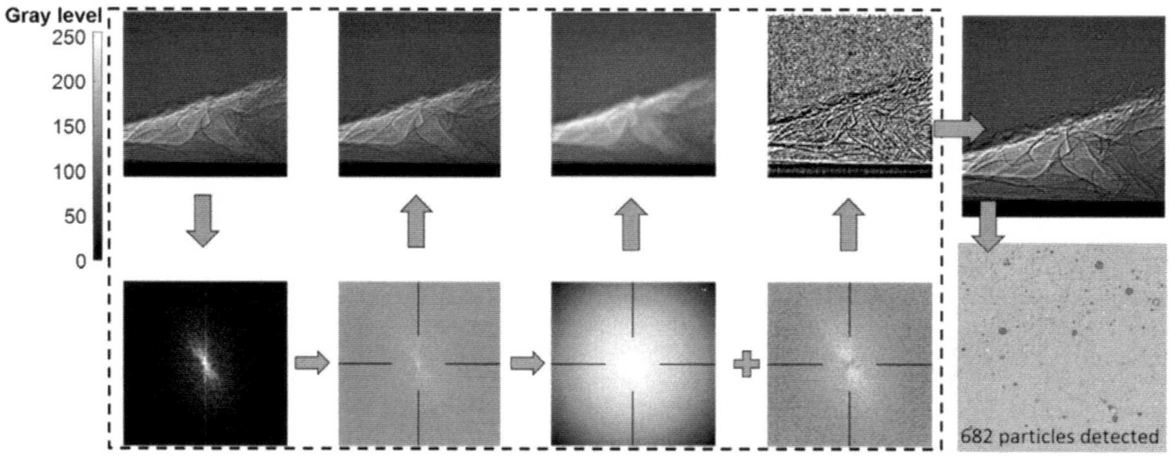

Figure 5: Edge detection on X-ray image with Gaussian filtering contrast enhancement.

Wavelet decomposition is commonly used to denoise two-dimensional signals, such as images. It can decompose an image into the approximation component and the detail component at different levels. The high-frequency signals like noise are included in the detail component and can be removed by proper thresholding. In this case, the intensity of particles varies relatively fast over space compared to the background of water and large-area vapor structures. Therefore, it is reasonable to treat extracting particles as denoising an image while performing a multi-level wavelet decomposition. Figure 6 shows the decomposition results by performing a 4-level discrete wavelet transform with the db5 wavelet function, where (a) is the approximation component and (d) is the detail component containing edge information of particles and vapor bubbles. Applying the edge detection algorithm on the detail component and 628 particles are detected.

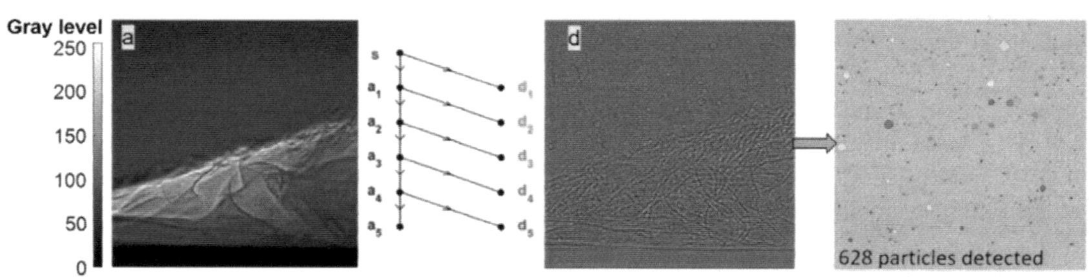

Figure 6: Edge detection on X-ray image with wavelet decomposition contrast enhancement.

All tracer particles detected from Figures 3 to 5 are combined and finally restored from the raw X-ray as shown in Figure 6 (left). The detected particle number is 1111 which is sufficient for the cross-correlation algorithm of the particle image velocimetry. An instantaneous velocity field of the liquid phase is shown, based on the pair of consecutive particle images with a time interval of 3.68 μs. Due to the thickness of the flow channel in this paper is 5 mm according to [16], the magnification uncertainty is a concern for researchers doing PIV measurement. With using X-ray imaging, the comparison between imposed and measured values shows that the mean error of the liquid phase velocity measurements is around ±0.46 m/s (~5% of the reference velocity of the studied flow [18]).

To obtain vapor tracers, the identified particles are removed from the initial image after intensity equalization and then replaced by the local average intensity of its neighborhood. Figure 6 (right) presents the final result of vapor structures which are used to calculate the instantaneous velocity field of the vapor phase. Since the vapor has a different x-ray attenuation coefficient from the liquid, the local vapor volume fraction α could be estimated quantitatively using Equation 1, derived from Lambert–Beer's law,

$$\alpha = 1 - \frac{ln(I_0/I_\alpha)}{ln(I_0/I_1)} \qquad (1)$$

where I_0 is the local intensity measured when the test section is filled with air, I_1 is the local intensity measured when the test section is full of water, and I_α is the local intensity measured when cavitation occurs in the test section. As shown in the plot, the maximum void fraction is around 85% locating at the interface between liquid and vapor zones, with the measurement uncertainty of about 2% [18].

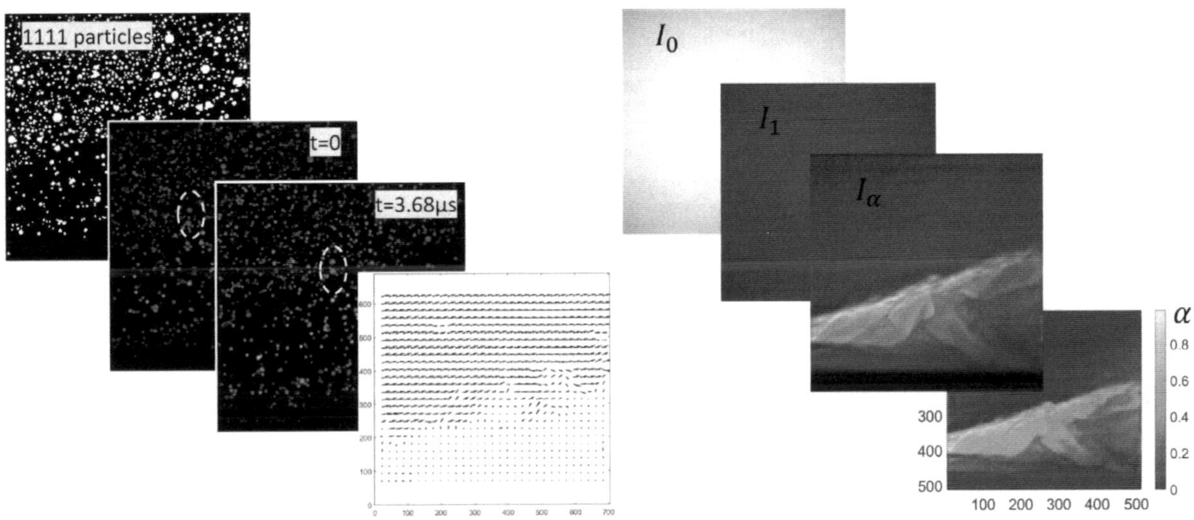

Figure 7: Sketch for the (left) particle velocimetry and (right) vapor void fraction calculation.

CONCLUSION

In this paper, the turbulent cavitating flow in a small size Venturi tube is investigated by the ultra-fast X-ray imaging technique. An original experiment based on X-ray imaging has been presented that the liquid tracer particles and vapor structures are both visualized clearly. A robust image processing algorithm has been developed for the simultaneous determination of the liquid velocities and void fractions in a cavitating flow. Images of radio-opaque particles and vapor bubbles have been recorded at high frequency, and post-processing has enabled us to obtain images of particles only or bubbles only, for PIV treatment. The local void fraction is estimated based on X-ray attenuation. Based on the acquisition of instantaneous velocity and void fraction fields with acceptable uncertainty, the analysis of the structures and dynamics inside and outside the cavity sheet can be investigated in our future works.

ACKNOWLEDGEMENTS

The authors would like to thank this project funded by ONR, under Grant No. N00014-18-S-B001. The use of the Advanced Photon Source at Argonne National Laboratory was supported by the U. S. Department of Energy, Office of Science, and Office of Basic Energy Sciences.

REFERENCES

[1] Dular, M., Khlifa, I., Fuzier, S., Maiga, M. A., & Coutier-Delgosha, O. (2012). Scale effect on unsteady cloud cavitation. Experiments in fluids, 53(5), 1233-1250.

[2] Wu, X., Maheux, E., & Chahine, G. L. (2017). An experimental study of sheet to cloud cavitation. Experimental Thermal and Fluid Science, 83, 129-140.

[3] Stutz, B., & Reboud, J. L. (1997). Experiments on unsteady cavitation. Experiments in fluids, 22(3), 191-198.

[4] Stutz, B., & Reboud, J. L. (1997). Two-phase flow structure of sheet cavitation. Physics of fluids, 9(12), 3678-3686.

[5] Stutz, B., & Reboud, J. L. (2000). Measurements within unsteady cavitation. Experiments in fluids, 29(6), 545-552.

[6] Gopalan, S., & Katz, J. (2000). Flow structure and modeling issues in the closure region of attached cavitation. Physics of fluids, 12(4), 895-911.

[7] Laberteaux, K. R., & Ceccio, S. L. (2001). Partial cavity flows. Part 1. Cavities forming on models without spanwise variation. Journal of Fluid Mechanics, 431, 1-41.

[8] Aeschlimann, V., Barre, S., & Djeridi, H. (2011). Velocity field analysis in an experimental cavitating mixing layer. Physics of Fluids, 23(5), 055105.

[9] Iyer, C. O., & Ceccio, S. L. (2002). The influence of developed cavitation on the flow of a turbulent shear layer. Physics of fluids, 14(10), 3414-3431.

[10] Coutier-Delgosha, O., Stutz, B., Vabre, A., & Legoupil, S. (2007). Analysis of cavitating flow structure by experimental and numerical investigations. Journal of Fluid Mechanics, 578, 171-222.

[11] Aeschlimann, V., Barre, S., & Legoupil, S. (2011). X-ray attenuation measurements in a cavitating mixing layer for instantaneous two-dimensional void ratio determination. Physics of Fluids, 23(5), 055101.

[12] Ganesh, H., Mäkiharju, S. A., & Ceccio, S. L. (2016). Bubbly shock propagation as a mechanism for sheet-to-cloud transition of partial cavities. Journal of Fluid Mechanics, 802, 37-78.

[13] Coutier-Delgosha, O., Vabre, A., Hocevar, M., Delion, R., Dazin, A., Lazaro, D., & Lee, W. K. (2009, January). Local Measurements in Cavitating Flow by Ultra-Fast X-Ray Imaging. In ASME 2009 Fluids Engineering Division Summer Meeting (pp. 371-379). American Society of Mechanical Engineers.

[14] Khlifa, I. & Coutier-Delgosha, O. (2013). Velocity measurements in cavitating flows using fast x-ray imaging. Congrés Français de Mécanique, France.

[15] Khlifa, I., Vabre, A., Marko Hočevar, Fezzaa, K., Fuzier, S., Roussette, O., & Coutier-Delgosha, O. (2017). Fast x-ray imaging of cavitating flows. Experiments in Fluids, 58(11), 157.

[16] Ge, M., Petkovšek, M., Zhang, G., Jacobs, D., & Coutier-Delgosha, O. (2021). Cavitation dynamics and thermodynamic effects at elevated temperatures in a small Venturi channel. International Journal of Heat and Mass Transfer, 170, 120970.

[17] Zhang, G., Khlifa, I., & Coutier-Delgosha, O. (2019, July). Experimental investigation of turbulent cavitating flows in a small venturi nozzle. In ASME-JSME-KSME 2019 8th Joint Fluids Engineering Conference. American Society of Mechanical Engineers Digital Collection.

[18] Zhang, G., Khlifa, I., Fezzaa, K., Ge, M., & Coutier-Delgosha, O. (2020). Experimental investigation of internal two-phase flow structures and dynamics of quasi-stable sheet cavitation by fast synchrotron x-ray imaging. Physics of Fluids, 32(11), 113310.

SNAME
THE INTERNATIONAL COMMUNITY
FOR MARITIME AND OCEAN PROFESSIONALS

MITIGATION OF INFECTIOUS DISEASE OUTBREAKS: CONSIDERATIONS FOR THE MARINE AND OFFSHORE INDUSTRIES

Martin Petričić
Senior engineer, Technology, American Bureau of Shipping (ABS)

Gareth Burton
Vice President, Technology, American Bureau of Shipping (ABS)

Libin Miao
Manager, Technology, American Bureau of Shipping (ABS)

Proceedings of the 26th Offshore Symposium, April 6th and 7th, 2021, Houston (Virtual), Texas
Texas Section of the Society of Naval Architects and Marine Engineers

ABSTRACT

Similar to land-based facilities, marine and offshore assets can be exposed to outbreaks of infectious diseases. This is exemplified by the recent COVID-19 outbreak which has had a significant impact on both personnel health and normal operations of the assets. The occurrence and transmission of infectious diseases on marine and offshore assets can, however, be mitigated by appropriate physical arrangements on board and having operational procedures in place. This paper addresses the former. The effectiveness of operational measures can be significantly increased if infectious diseases are considered at the asset's design stage. ABS has identified a lack of available technical guidance regarding physical arrangements that can help make marine and offshore assets safer for crews and reduce the spread of infectious diseases, such as COVD-19. This paper presents the best practices and recommendations for physical arrangements drawn mainly from the recognized standards for land-based medical facilities and applied to marine and offshore assets. The recommendations focus on specific spaces inside the accommodation block, their number, location, layout, ventilation, and interior surfaces. Isolation cabins with their associated anterooms are proposed as the most reliable way of isolating suspected or confirmed cases of an infectious disease from the rest of the personnel on board. Negative pressure, independent exhaust system, and easy to clean surfaces that are accessible and resistant to deterioration from frequent cleaning and disinfection are effective measures in containing the transmission of infectious diseases that are spread through the air or direct and indirect contact. The paper also emphasizes the need to promote segregation of the visitors from crew and passengers by providing designated spaces for the visitors and gives recommendations on the communication equipment needed for receiving medical advice from land-based medical specialists. If properly implemented and if augmented with robust operational measures, physical arrangement measures have the potential to significantly improve the safety of the crew, passengers, and visitors, as well as minimize the negative consequences of disruptions to normal asset operations.

Keywords: infectious diseases, mitigation, physical arrangement, isolation cabin, anteroom, ventilation, negative pressure, interior surfaces, telemedical assistance, marine, offshore

INTRODUCTION

The impact of the global COVID-19 pandemic has highlighted how marine and offshore assets can be exposed to outbreaks of infectious diseases just as readily as land-based facilities. Outbreaks have a significant impact on both personnel health and normal operations of the assets.

Where infectious disease risks occur, it is very important to implement robust physical arrangement measures and operational measures. Both sets of measures can reduce the risk of contracting and transmitting infectious disease, thus protecting the health of the crew, passengers, and others in contact with the asset, while maintaining the asset in an operational state.

Each infectious disease outbreak is different and may present different operational challenges. Also, the same objectives of early detection, prevention, and control of infectious diseases may be achieved using many different operational measures, which are outside the scope of this paper. Guidelines on the operational measures to mitigate the risks of COVID-19 and other similar infectious diseases that are transmitted via air or direct and indirect contact can be found in the ABS *Guidance Notes on Response Measures to COVID-19 for the Marine and Offshore Industries* [1]. This paper only addresses the physical arrangement measures such as layout and location of accommodation and work spaces, ventilation system design, interior surfaces, and the necessary equipment to be installed on board.

Incorporating physical arrangement measures for infectious disease mitigation at the design stage can significantly improve the success of subsequent operational measures taken on board marine and offshore assets.

In order to fill in the gaps in existing regulations and standards, ABS has presented the marine and offshore industry with guidance on addressing the technical aspects of an asset's physical arrangements to mitigate infectious disease outbreaks. This guidance has been developed based on consultations with medical experts and several independent governmental and commercial sources including the United States Centers for Disease Control and Prevention (CDC) [2], American National Standards Institute (ANSI) [3], [4], American Society of Heating, Refrigerating and Air-Conditioning Engineers (ASHRAE) [3], [4], American Society for Healthcare Engineering (ASHE) [3], The Facility Guidelines Institute [5], The American Institute of Architects (AIA) [5], the International Organization for Standardization (ISO) [6], [7], and the World Health Organization (WHO) [8], [9], [10]. In addition, ABS is now offering an optional notation, Infectious Disease Mitigation-Arrangements (IDM-A) [11].

INFECTIOUS DISEASE TRANSMISSION

Many infectious diseases are caused and spread by microbes including viruses, bacteria, fungi, and protozoa. Some infectious agents can remain viable in air, water, food, on surfaces, and in nearly all bodily fluids. Many infectious agents remain viable long enough to enable an indirect transfer from one person to another via one of these substrates, which can cause or increase an outbreak.

Given the environment on marine and offshore assets is a restricted one, infectious diseases have the potential to spread rapidly and affect significant proportions of the personnel on board.

On marine and offshore assets, infectious diseases are transmitted through six main routes:

- Food
- Water
- Vectors (e.g., insects and rodents)
- Air – droplet or aerosol transmission
- Direct contact with an infected person
- Indirect contact with contaminated objects and surfaces

When the infectious disease is transmitted through the air, the infectious agents may be contained in either droplets or droplet nuclei (aerosol). Droplets are greater than 5 μm in size and are usually generated when a person coughs or sneezes. The droplets containing infectious agents can then reach another person who is standing in proximity (usually within a few meters). This can constitute a direct form of infectious disease transmission. The droplets tend to fall out of the air quickly and do not travel large distances. Droplet nuclei are the residue of droplets. They are far smaller, usually between 1 μm and 5 μm in size, and can remain suspended in the air for a very long time. Droplet nuclei can travel large distances and can constitute an indirect form of infectious disease transmission.

EXISTING REGULATORY REQUIREMENTS AND STANDARDS

Numerous regulatory bodies have issued standards for general sanitary conditions of marine and offshore assets. For example, the International Health Regulations [8] governed by the World Health Organization offer very good requirements and best practices regarding food and water safety, vector control, and waste management on marine vessels on international voyages.

However, following an extensive literature review and consultations with prominent medical experts, ABS has identified a lack of explicit technical guidance regarding physical arrangements to help make marine and offshore assets more resistant to the occurrence and transmission of infectious diseases such as COVID-19, whether they are spread via air or direct and indirect forms of contact.

PHYSICAL ARRANGEMENT MEASURES

ISOLATION CABINS

The most important physical arrangement feature of an asset, intended to mitigate the risk of infectious disease transmission, is the existence of dedicated or convertible regular single-occupancy isolation cabins that are served by an anteroom and have an independent sanitary space with a toilet and a shower. Isolation cabins serve to isolate a suspected or confirmed case of an infectious disease as is commonly done in land-based hospitals. Single-occupancy isolation cabins should be capable of holding at least 5% of the asset's total complement. The 5% threshold delineates between a moderate and a major outbreak on board an asset.

Anterooms are spaces directly connected to the isolation cabin and are used for personnel entering the isolation cabins to don (put on) and doff (take off) personal protective equipment (PPE). They serve to protect the people entering the isolation cabins and to enhance the prevention of infectious disease transmission.

The isolation cabins can be provided inside the accommodation block, or as portable accommodation modules installed on the open deck space. If provided inside the accommodation block, they should be located towards the end of an accommodation corridor, close to a direct exit to the outside space. This is to minimize the exposure of accommodation block spaces to infected individuals while they are being disembarked from the asset.

Figure 1 shows possible arrangements of the isolation cabins and anterooms inside the accommodation block. More than one isolation cabin can be served by a common anteroom (see Figure 1 b and d). If two adjacent cabins share a sanitary space and one of them is used as an isolation cabin, while the other one is used as an anteroom (see Figure 1 c), then the shared sanitary space can typically be used as a passageway. In this arrangement, the shared sanitary space is a part of the isolation cabin. A corridor can also be used as an anteroom serving multiple isolation cabins (see Figure 1 d). Because access to the anteroom should be restricted during an outbreak, a part of the corridor serving as an anteroom cannot be a part of any escape route, other than for the occupants of the isolation cabins and the anteroom.

In order to minimize the risk of contaminated aerosol from the isolation cabin or anteroom spreading to other areas of the accommodation block, the isolation cabin should be under at least 2.5 Pa of negative pressure with respect to the anteroom, and the anteroom should be under at least 2.5 Pa of negative pressure with respect to the adjacent corridor. The negative pressure can be achieved by making the exhaust airflow greater than the supply airflow. In addition to this, isolation cabins and anterooms should be well ventilated using mechanical ventilation with at least 12 air changes per hour (ACH) for each isolation cabin, and 10 ACH for the anterooms.

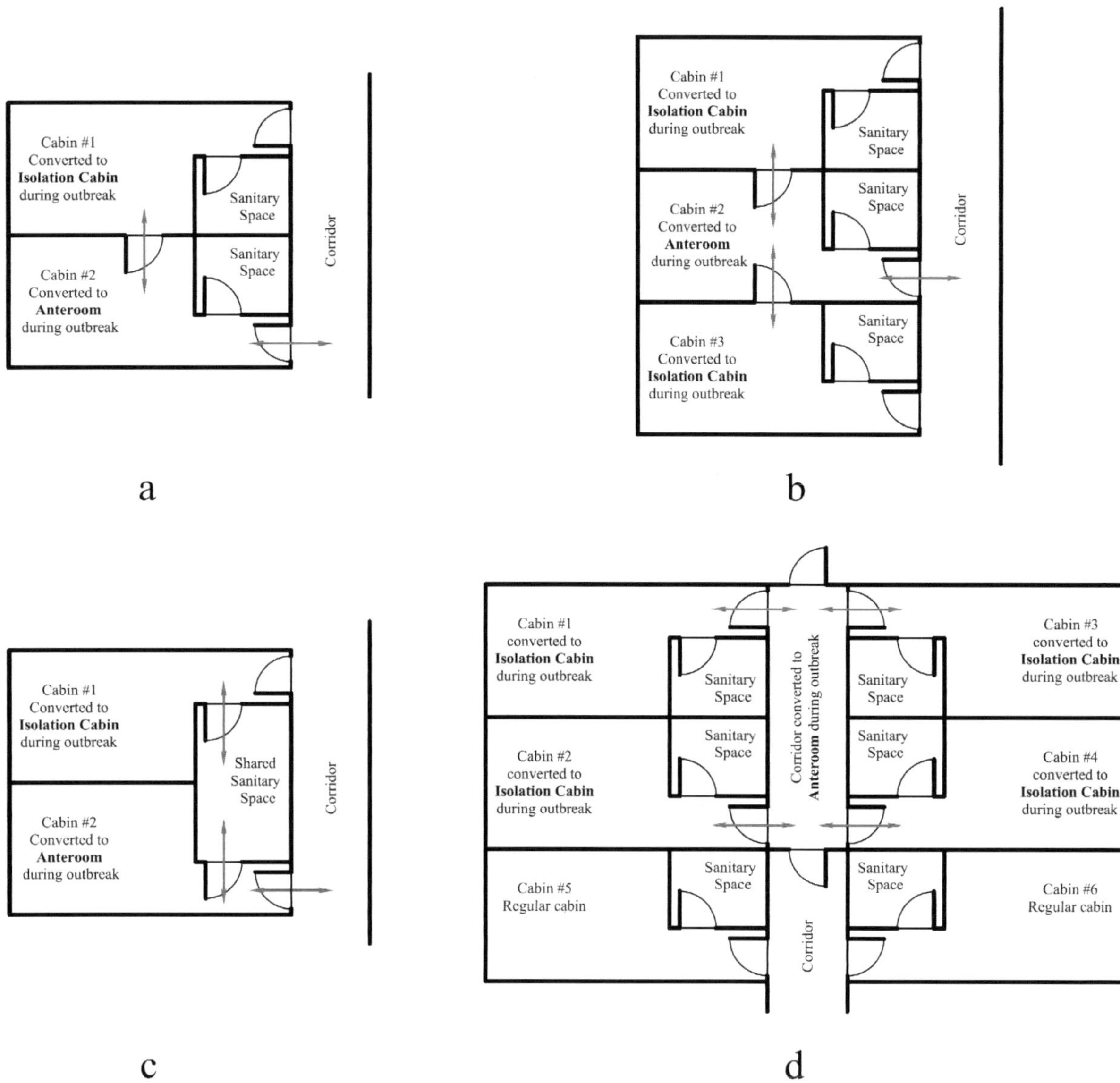

Figure 1: Different layout options for isolation cabins and anterooms

Attention should also be paid to the isolation cabin and anteroom door swing direction. In general, the doors should always swing into the area of higher pressure so that the net pressure on the doors acts to close the doors, if left opened. The doors should also have a seal around their edges and a self-closing mechanism without a holdback device. Any ventilation louvers on the doors should be completely closed during an outbreak. Access to the isolation cabins and anterooms should be restricted during the outbreak. In order to enable emergency egress from the isolation cabins and anterooms, their doors should not require a key to unlock from the inside.

All major air leaks from the isolation cabins and anterooms should be minimized by sealing all openings, penetrations, and bulkhead-ceiling interface, using monolithic ceilings, and coving up the floor coverings up the bulkheads. Minimizing the air leakage minimizes the risk of aerosols escaping from the isolation cabins and anterooms, especially if the negative pressure system fails. This also minimizes the load on the exhaust fan which is used to create the negative pressure.

The exhaust air duct from the isolation cabins and anterooms should be completely separated from all other exhaust ducts in order to minimize the risk of accidental contamination of other areas of the accommodation block. However, if the exhausted air from each isolation cabin and anteroom is filtered using a high-efficiency particulate air (HEPA) filter, then the air from the isolation cabins and anterooms can be considered as non-contaminated and can be mixed with the exhaust air from other spaces, and even recirculated [5], [2]. The exhaust air drawn directly from sanitary spaces inside the isolation cabins and anterooms should not be recirculated, but if it is HEPA filtered, it can be mixed with non-recirculated exhaust air from other spaces.

The supply air to the isolation cabins and anterooms should be filtered using a 90% dust spot efficiency filter. If more than one isolation cabin and anteroom are connected to the same supply air duct or the same exhaust air duct, then each local supply or exhaust duct branch to each isolation cabin or anteroom should have a gas-tight, non-return damper (see Figure 2 a). This will minimize the risk of isolation cabin and anteroom cross-contamination, especially in the event of ventilation system failure.

Gas-tight, non-return dampers may be substituted with HEPA filters. One advantage of using HEPA filters to filter the exhaust air is that no special consideration of the exhaust outlet location is needed. Otherwise, the exhaust outlet should be at least 8 meters away from any air intakes, natural ventilation points, doors, and open-type windows. Figure 2 b shows an alternative exhaust air arrangement where the gas-tight, non-return valves are replaced with HEPA filters. This enables the air from the isolation cabins and anterooms, excluding the air from their sanitary spaces, to be re-circulated. It also enables the exhaust air from isolation cabin and anteroom sanitary spaces to be mixed with the exhaust from other sanitary spaces and exhausted outside through the normal exhaust outlet.

HEPA filters can offer a 99.97% efficiency rate for filtering particles larger than 0.3 μm. Aerosols are typically 1 μm to 5 μm in size and are, thus, significantly larger than the HEPA filter particle threshold. The Facility Guidelines Institute and The American Institute of Architects [5] as well as the CDC [2] recommend HEPA filtration of the exhaust air from the hospital Airborne Infection Isolation Rooms (AIIRs) prior to recirculation. Ultraviolet germicidal irradiation (UVGI) lamps are less effective at higher duct air speeds [12], [13] and should not be used instead of HEPA filters [2]. However, they may be used in conjunction with HEPA filters.

MEDICAL FACILITY

In order to minimize the exposure of accommodation block spaces to infected individuals while they are being brought to or escorted from the medical facility, attention should be paid to the location of the medical facility on board the asset. The asset's medical facility or hospital should be located as close as possible to the entrance into the accommodation block from an open deck space.

The ventilation system of the medical facility should be similar to the ventilation system of isolation cabins. The medical facility should be under negative pressure with an independent exhaust duct to the outside. Similar door and sealing considerations for the isolation cabins and anterooms should be applied to a medical facility.

DEDICATED SANITARY AND OFFICE SPACES FOR VISITORS AND SHORE PERSONNEL

In order to promote social distancing between the crew/passengers and visitors, designated sanitary and office spaces for visitors, should be considered at the design stage of an asset. During an outbreak, these spaces should be used exclusively by shore personnel and other visitors while conducting the necessary business on board an asset, thus enabling normal asset operations while minimizing the risk of infectious disease transmission.

In order to further minimize the exposure of crew/passengers and accommodation block spaces to potentially infected visitors, designated sanitary and office spaces for visitors should be readily accessible from an outside entrance into the accommodation block. These spaces can also be provided as portable modules on an open deck space or as existing accommodation block office and sanitary spaces that can be converted into the designated office and sanitary spaces.

The designated sanitary and office spaces for visitors are to be well ventilated using mechanical ventilation with independent exhaust to the outside air. If the exhaust air is first HEPA filtered, it can then be mixed with other exhausts and without a special consideration of the outlet location. HEPA filtered exhaust air from the designated visitor office space can be recirculated.

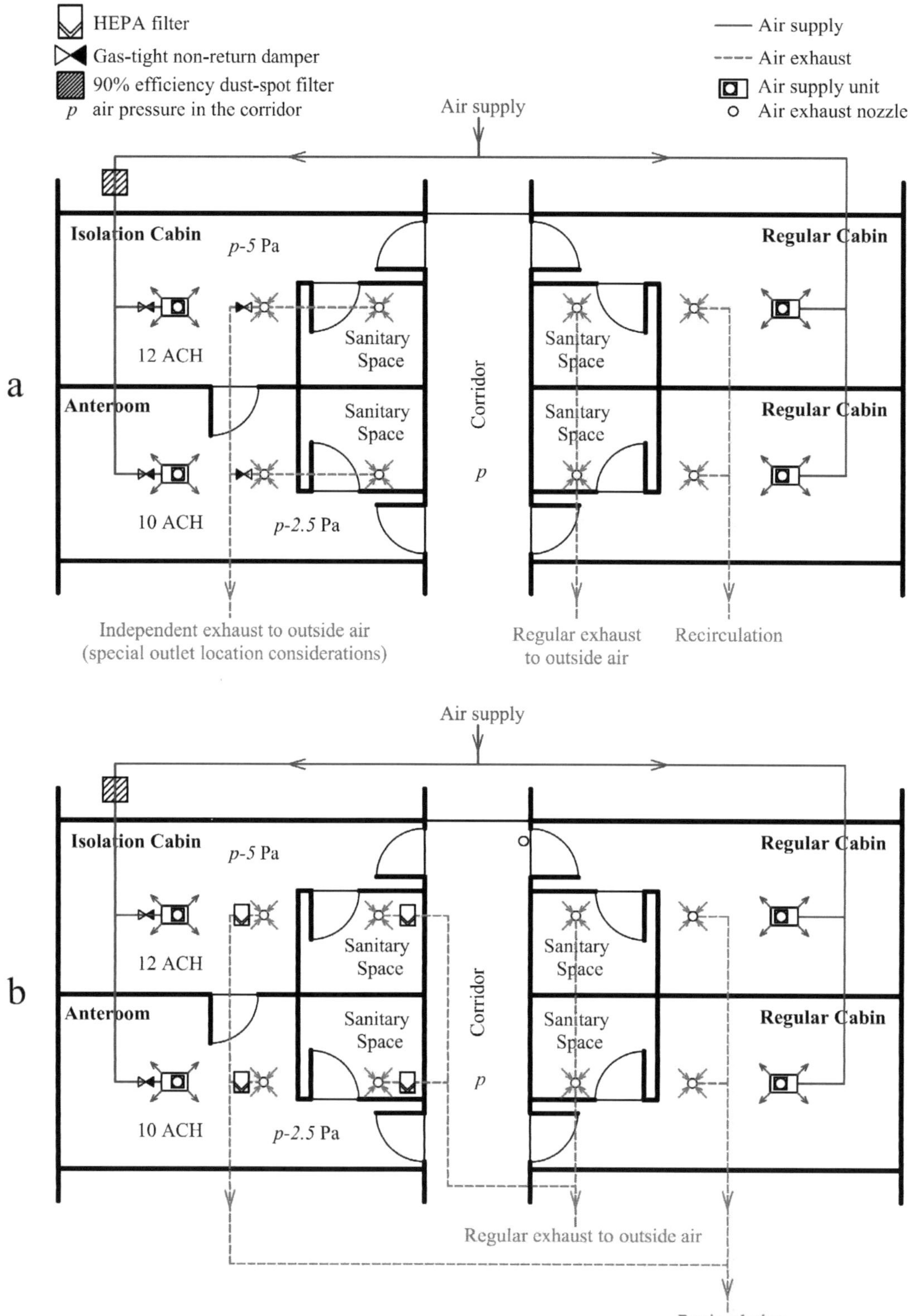

Figure 2: Isolation cabin and anteroom ventilation system arrangement alternatives

Storeroom for Infectious Solid Waste

During the outbreak of an infectious disease, contaminated infectious waste is generated on board an asset and needs to be disposed of. While the liquid infectious waste can be safely disposed of through the asset's sewage system and sewage treatment plant, if available, the solid infectious waste should be disposed of ashore. Before disposal ashore, the solid infectious waste should be collected in properly labeled and tightly closed plastic bags and put into a designated storeroom for solid infectious waste. This space should be separated from all other spaces with a properly labeled entrance which can be from a regular storage space for non-infectious waste. Alternatively, this space can also be provided as a portable module on an open deck space. In order to minimize the proliferation of microbes, direct sunlight should be blocked by shades or blinds on all windows.

The storeroom for solid infectious waste should be well ventilated using mechanical ventilation with independent exhaust to the outside air. Similarly, as for other spaces with potentially contaminated air, HEPA filters can be used to enable mixing the exhaust with other non-contaminated exhausts and to avoid a special consideration of the outlet location. The air from the storeroom for solid infectious waste should not be recirculated.

Storeroom for Cleaning Agents and Disinfectants

During an outbreak on board an asset, or during an epidemic in a local area in which the asset is located, there will be a need for continuous and frequent cleaning and disinfection of many spaces and items on board marine and offshore assets. Therefore, adequately ventilated and sun-shaded spaces should be provided on board an asset to store cleaning agents and disinfectants.

Special attention is to be paid to separating oxidizing agents such as bleach (calcium hypochlorite, sodium hypochlorite) or hydrogen peroxide and flammable agents such as alcohol-based disinfectants. When stored together, vapors from oxidizing and flammable agents can create a highly combustible atmosphere. Therefore, two separate storerooms should be provided for storing these two groups of agents.

Interior Surfaces

The interior surfaces inside many of the aforementioned accommodation spaces should be carefully designed in terms of their material and surface properties. Regarding mitigation of infectious disease transmission, the properties that the interior surface should have are:

1. Resistance to the accumulation of dirt that can harbor dangerous microbes
   - Moisture impervious surfaces
   - Free of micro-perforations, corrugations, and significant roughness
2. Easy to clean and disinfect
   - All exposed surfaces including joiner panels and deck coverings should be accessible for cleaning and disinfection
   - Where possible, interior corners are to have a radius to facilitate cleaning
   - Textiles and fabrics that are difficult to launder should be avoided on floors, bulkheads, and furniture
   - Mattresses should have impermeable covers
3. Resistance to the degradation from frequent cleaning and disinfection
   - Bare metal surfaces, other than stainless steel and galvanized steel, are to be avoided as they can be exposed to chlorine-based stress corrosion cracking
   - Untreated wooden surfaces without a finish (paint, lacquer, varnish, drying oil, wax, or shellac) are to be avoided as they are not moisture impervious and will not be resistant to the application of cleaning and disinfecting agents
   - Coatings on interior surfaces should be free of any visible cracks or defects

There are many coatings for different types of surfaces being currently developed that claim to have antimicrobial properties. However, their effectiveness and suitability for application on marine and offshore assets have not been sufficiently studied.

Additional Equipment

Outbreaks of infectious diseases can spread rapidly on board marine and offshore assets and can quickly overwhelm the crew's capacity to properly handle the situation, even if medical doctors are present on board. Good preparation and appropriate procedures and protocols can mitigate the outbreak, but it is also very important that the asset has the appropriate communication equipment so that medical advice can be provided to onboard doctors or medical care personnel from land-based medical facilities that have specialists on constant duty. One such telemedical service is provided, free-of-charge, by the Telemedical Assistance Service (TMAS) [14] to all marine and offshore assets around the world. The communication equipment installed on board the asset should, at a minimum, allow for two-way voice communication with TMAS or other similar services from anywhere within the asset's service area. In addition to the voice communication, it is also recommended that the asset has the capability to transmit data to TMAS, such as data from the medical diagnostic equipment, photographs, or videos.

Examples of communication equipment for establishing communication with telemedical service are:

- Very High Frequency (VHF) radio – has a limited range
- High Frequency (HF) radio – has a very long range
- Inmarsat communications - abridged Inmarsat system code 32 is used to obtain medical advice
- Satellite telephone
- Satellite internet connection – allows for voice and data/video communications

Other equipment installed on board that may aid in mitigating the outbreaks of infectious diseases include:

- Washing machine capable of a bleach cycle with a water temperature of at least 71°C (160°F). Hot water of this temperature provides an effective means of destroying microorganisms [2]. The bleach will aid in sanitizing the contaminated towels, linens, and other washable fabrics and will provide an extra margin of safety.
- Touchless doors, taps, toilets, soap, and hand sanitizer dispensers, elevators, etc. Touchless technology reduces the number of frequently touched surfaces and reduces the risk of infectious disease transmission via indirect contact.
- Hypochlorous acid (electrolyzed water) generators. Hypochlorous acid is a powerful disinfectant that can be generated by water electrolysis on board an asset.

In addition to the aforementioned equipment, the following technology also exists that may aid in mitigating the outbreaks of infectious diseases:

- Active air purification systems for marine applications
- Radio Frequency Identification (RFID) tags that can be used for contact tracing, access control, traffic flow control, and hands-free access

These technologies do not have a significant application track record in the marine and offshore industries and require more testing and validation of their effectiveness.

Conclusion

Given the restricted environment on marine and offshore assets, infectious diseases have the potential to spread rapidly and can have a very negative impact on the health and safety of the crew and passengers, as well as on the normal operations of the asset. The COVID-19 pandemic has exposed a lack of available technical guidance regarding physical arrangements that can help make marine and offshore assets safer for crews and reduce the spread of infectious diseases that are transmitted via air or direct and indirect forms of contact.

Following an extensive literature review of existing standards and regulations from various national and international health and engineering organizations, as well as consultations with medical personnel, ABS has identified the aspects of physical arrangements that are likely to be most effective in mitigating the occurrence and transmission of infectious diseases on board. These measures focus on specific spaces inside the accommodation block, their number, location, layout, ventilation, and interior surfaces. Isolation cabins with their associated anterooms are proposed as a means of isolating suspected or confirmed cases of an infectious disease from the rest of the personnel on board. Special considerations should be given to ventilation specifications of these cabins to mitigate the spread of airborne viruses such as negative pressure, sufficient air changes, and treatment of the exhaust air. Other physical arrangement measures include the location and ventilation specifications of medical facilities on board, providing designated sanitary and office spaces for visitors to promote personnel and visitor segregation, storerooms for infectious solid waste, and telemedicine capabilities. Easy to clean surfaces that are accessible and resistant to deterioration from frequent cleaning and disinfection are also effective measures in containing the spread of infectious diseases that are spread through direct and indirect contact.

Physical arrangement measures have the potential to significantly increase the effectiveness of the operational measures that are implemented before and during the outbreak. The most cost-effective way is to implement the physical arrangement measures at the design stage of the asset.

If properly implemented and if augmented with robust operational measures, physical arrangement measures have the potential to improve the safety of the crew, passengers, and visitors as well as to minimize the negative consequences of disruptions to normal asset operations.

ACKNOWLEDGEMENTS

The authors would like to acknowledge the support of epidemiologist Joyce Johnson, Rear Admiral DO MA U.S. Public Health Service (Ret), in developing this guidance.

REFERENCES

[1] ABS, Guidance Notes on Response Measures to COVID-19 for the Marine and Offshore Industries.

[2] CDC, "Guidelines for Environmental Infection Control in Health-Care Facilities," July 2019. [Online]. Available: https://www.cdc.gov/infectioncontrol/guidelines/environmental/index.html. [Accessed February 2021].

[3] ANSI/ASHRAE/ASHE, "Standard 170-2008: Ventilation of Health Care Facilities," American Society of Heating, Refrigerating and Air-Conditioning Engineers, Atlanta, 2008.

[4] ANSI/ASHRAE, "Standard 55-2013: Thermal Environmental Conditions for Human Occupancy," American National Standards Institute, Atlanta, GA, 2013.

[5] FGI, AIA, "Guidelines for Design and Construction of Hospital and Health Care Facilities," The American Institute of Architects, Washington, 2006.

[6] ISO, "Standard 7547:2002 Ships and marine technology — Air-conditioning and ventilation of accommodation spaces — Design conditions and basis of calculations," International Organization for Standardization, Geneva., 2002.

[7] ISO, "Standard 7726:1998(E) Ergonomics of the thermal environment – Instruments for measuring physical quantities," International Organization for Standardization, Geneva, 1998.

[8] WHO, International Health Regulations (2005), 2nd ed., Geneva: World Health Organization, 2008.

[9] WHO, "Handbook for Inspection of Ships and Issuance of Ship Sanitation Certif cates," World Health Organization, Geneva, 2011.

[10] WHO, "Guide to ship sanitation," World Health Organization, Geneva, 2011.

[11] ABS, Guide for Mitigation of Infectious Disease Transmission on Board Marine and Offshore Assets.

[12] Y. Yang, H. Zhang, S. S. Nunayon, V. Chan and A. C. K. Lai, "Disinfection Efficacy of Ultraviolet Germicidal Irradiation on Airborne Bacteria in Ventilation Ducts," *Indoor Air,* vol. 28, no. 6, p. 806–817, 2018.

[13] E. S. Mousavi, N. Kananizadeh, R. A. Martinello and J. D. Sherman, "COVID-19 Outbreak and Hospital Air Quality: A Systematic Review of Evidence on Air Filtration and Recirculation," *Environmental Science & Technology,* 2020.

[14] IMO, "MSC/Circ.960 Medical Assistance at Sea," International Maritime Organization, London, 2000.

THE INTERNATIONAL COMMUNITY
FOR MARITIME AND OCEAN PROFESSIONALS

LIFTING OPERATIONS WITH WIND TURBINE INSTALLATION JACK-UPS USING REAL-TIME DATA

KAYO VANDERHEGGEN, JOOST JANSSEN, NATE MEREDITH
GUSTOMSC | NOV

Proceedings of the 26th Offshore Symposium, April 6 & 7th 2021, Houston, Texas
Texas Section of the Society of Naval Architects and Marine Engineers

ABSTRACT

When a wind turbine installation jack-up performs a heavy lifting operation with the crane it affects the loads on the foundation. For these units the crane typically encircles a leg or is positioned close to it. Consequently, that leg attracts most of the loads due to crane operations.

For each location jack-ups prove the capacity of the foundation by applying a controlled, high load at each of the footings before commencing operations. This process is known as preloading. The achieved preload at the jack-up's foundation determines the operational limit. Exceedance of the preload value may result in foundation instability. Depending on the site's foundation characteristics the consequences of such an exceedance range from negligible to catastrophic failure.

GustoMSC has developed Operator Support System (OSS) software with the purpose to make the operator aware of the limitations imposed by the preloaded foundation. The application outlines operational limits based on real-time data from the jack-up, jacking system and crane which enables the operators to safely unlock the full potential of their wind turbine installation jack-up.

Keywords: wind turbine installation jack-up, crane operations, preload, load, foundation stability, reliability, measurement, big data, applications, Operator Support System

INTRODUCTION

For the installation of modern wind turbines dedicated jack-ups are used with specialized cranes. With the introduction of these highly specialized installation solutions, crane load induced foundation failure has turned from a statistically insignificant failure mechanism into one of the largest risks associated with wind turbine installation. The level of safety and performance of wind turbine installation jack-ups is currently suboptimal because clearly defined limits and target reliabilities for these critical operations do not yet exist.

When a wind turbine installation jack-up performs a heavy lift operation with the crane it affects the loads on the foundation. For these jack-ups the crane typically encircles a leg or is positioned close to it. Consequently, that leg attracts most of the loads during crane operations and therefore the foundation capacity under that leg can become a critical factor.

For each location, jack-ups prove the capacity of the foundation by applying a controlled, high load at each of the footings before commencing operations. This process is known as preloading. The achieved preload at the jack-up's foundation determines the operational limit. Exceedance of the preload value may result in foundation instability. Such a failure mechanism can be caused by a heavy load being rotated outboard as illustrated in Figure 1.

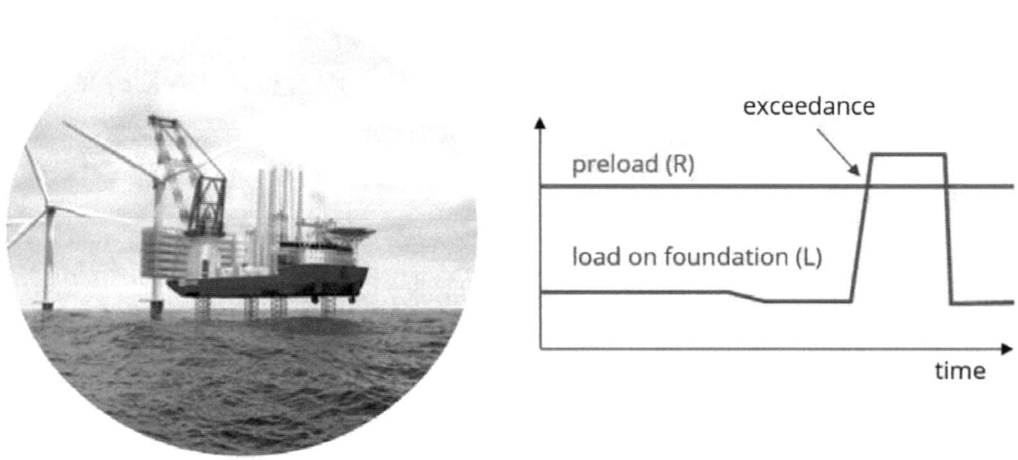

Figure 1: Foundation failure during lifting operation

Depending on the site's soil characteristics the consequences of such an exceedance can range from negligible to catastrophic failure and this failure can occur during lifting operations with *any* wind turbine installation jack-up. It all depends on the established preload and the size of the lift.

There is a clear need for understanding the risks associated with crane operations, establish healthy safety margins and align design assumptions with operator safety awareness. Warning signs indicating potential danger are not trivial for jack-ups. Specialized load measurement instrumentation is typically embedded in the jacking systems to visualize the loads.

Human judgement and operator experience is heavily relied upon to prevent accidents. Target reliabilities and safety factors for crane operational conditions have not yet been established. Consequently, there is no consistent industry wide basis for evaluating the acceptability of lifting operations for jack-ups in elevated conditions. That leads to situations where operators are unaware of limits, resulting in safety risks. On the other hand, it also leads to uneconomic use of equipment. Due to the uncertainty, conservatism tends to creep into the designs and operations.

This paper addresses the gaps that, from the perspective of the designer, prevent optimal and safe use of wind turbine installation jack-ups.

STRUCTURAL RELIABILITY BASICS

Structural engineers are used to dealing with reliability problems in a specific form, the verification of limit states using a single safety factor (working stress design) or partial safety factors (load and resistance factor design). This way of verifying reliability problems is of course very practical as a tool. It simplifies a statistical problem into a simple algebraic equation. The factored load should be less than the factored resistance:

$$\gamma_L F_L \leq \gamma_R F_R$$

For the engineer, this approach simplifies all the underlying assumptions and statistical considerations that go with the reliability problems. It is easy to forget that these safety factors are a result of a process where risks were evaluated, target reliabilities agreed upon, limit state functions defined, and of which the variables were modeled stochastically for numerous cases.

The rational approach to structural reliability starts not with a safety factor, but with setting a target probability of failure. This acceptable probability of failure depends on the risk. If the risk is high, one should aim for a low probability of failure. If the risk is low, one can settle for a higher probability of failure.

Once the target reliability is set, failure needs to be defined. The mathematical description of failure is done with limit state functions. These functions consist of stochastic variables. The most basic limit function has only two variables, a load and a resistance variable, represented by the equation:

$$m = r - l$$

Where m is the margin, r is the resistance and l is the load. The essence of the reliability problem is to ensure sufficient positive margin between the average load μ_L and average resistance μ_R, accounting for the statistical uncertainties related to these variables, their standard deviations σ_L and σ_R, such that the probability of failure P_f (when margin is less than zero), is sufficiently low. This is illustrated in Figure 2 assuming load and resistance variables are normally distributed functions.

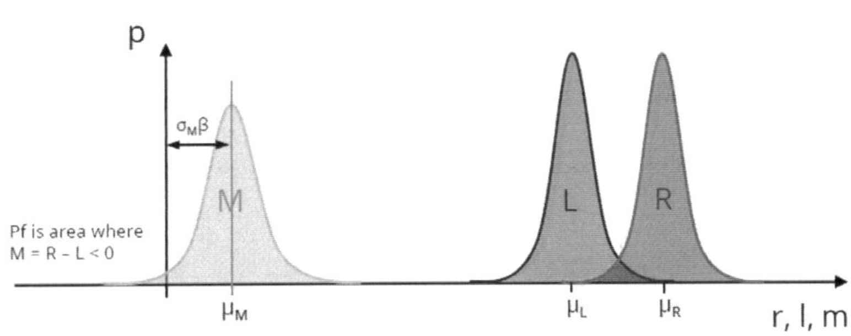

Figure 2: Distributions of resistance, load and margin

In the figure above the standard deviations for load and resistance are equal for simplicity. The parameters of the probability distribution of the margin are in general:

$$\mu_m = \mu_r - \mu_l$$

$$\sigma_m = \sqrt{\sigma_r^2 + \sigma_l^2}$$

The probability of failure P_f is the area where the margin is less than zero:

$$P_f = \int_{-\infty}^{0} p_m dm$$

Alternatively, the probability of failure can be represented as the reliability index, β:

$$\beta = \frac{\mu_m}{\sigma_m}$$

The safety factor is the practical result determined by the target probability level, or target reliability index, and the statistical uncertainties of the stochastic variables. Using this method, the process of calculating safety factors for crane operational conditions would comprise of the following steps:

1. Set a target acceptable probability of failure considering the risks
2. Determine limit state functions, the functions that mathematically describes failure
3. Identify the stochastic variables and determine their stochastic properties
4. Derive the resulting safety factor

Now, let's apply these steps to wind turbine installation jack-ups.

STRUCTURAL RELIABILITY OF WIND TURBINE INSTALLATION JACK-UPS

Wind turbine installation jack-up operators often optimize the preload, i.e. apply no more preload than required to safely perform the installation of a turbine and survive a storm. There are many reasons to do so: reduce wear and tear, avoid deeper penetrations, save time, and avoid prestressing the leg's truss structure during the preload operation induced by an uneven seabed. To do so safely, operators need a safe margin between the red line, the achieved preload, and the top of the blue line, the load on the foundation during a crane operation, in Figure 3. This figure represents the more desirable version of Figure 1: safe, but economically designed. In the following sections the basics of structural reliability are applied to quantify the safety margin.

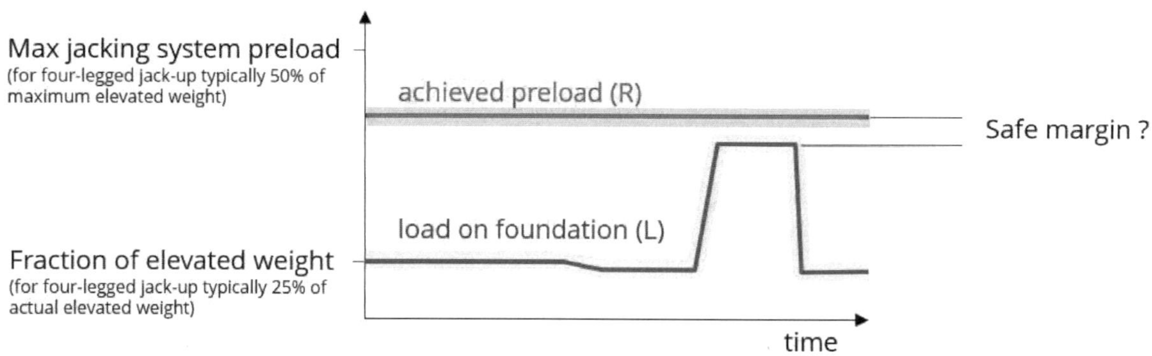

Figure 3: Safely optimizing the preload

Target probability of failure

To set a reasonable probability of failure, one must first analyze the consequences and quantify the risks. For the scope of this paper, failure is defined as the event where the load on the jack-up's foundation is exceeded (Figure 1). The consequences of failure depend on the foundation and the jack-up. The jack-up itself is important because some designs are more robust than others. But what is a typical wind turbine installation jack-up design?

According to the Global Wind Energy Council's database there are 137 wind turbine installation vessels worldwide, of which 82 are jack-up vessels that have participated in offshore wind turbine installation work [1]. Figure 4 breaks down GustoMSC's incomplete database of construction jack-ups based on number of legs, leg type and leg shape. The majority of the construction jack-ups have four legs and this large group may be split in two distinct groups of approximately even size: a truss leg and closed leg group. Figure 5 shows that the truss leg group typically uses a rack and pinion jacking system and electric motors as movers, while the closed leg group often uses a pin and hole jacking system driven by hydraulic pistons.

The preload capabilities of the wind turbine installation jack-ups range from 1,000 t to beyond 20,000 t and the lifting capacity of the offshore cranes ranges from 600 t to beyond 3,000 t. Older units typically use pedestal cranes, while more modern wind turbine installation jack-ups use leg encircling cranes.

Table 1 presents the basic properties of 5 units types that that give some perspective on the total group. The majority of the existing construction jack-ups resemble unit 2 and 3 from the table. The subsequent generation of wind turbine installation jack-ups resemble unit types 1 and 4, while the new generation of dedicated wind turbine installation jack-ups are similar to type 5. At least four jack-ups currently under construction are of type 5.

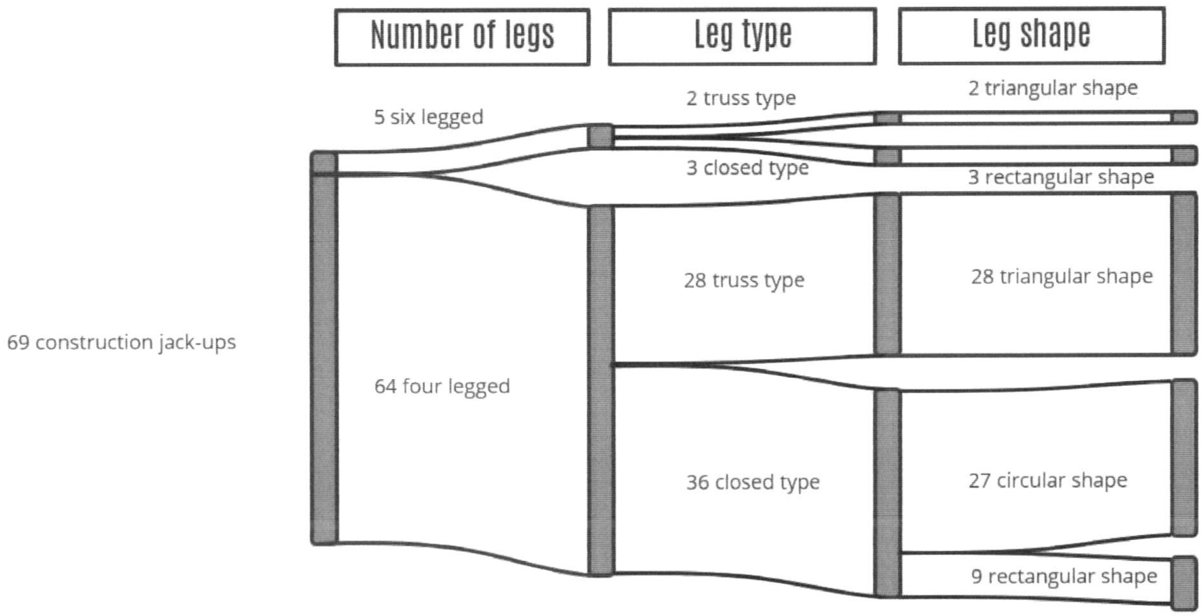

Figure 4: The construction jack-up landscape

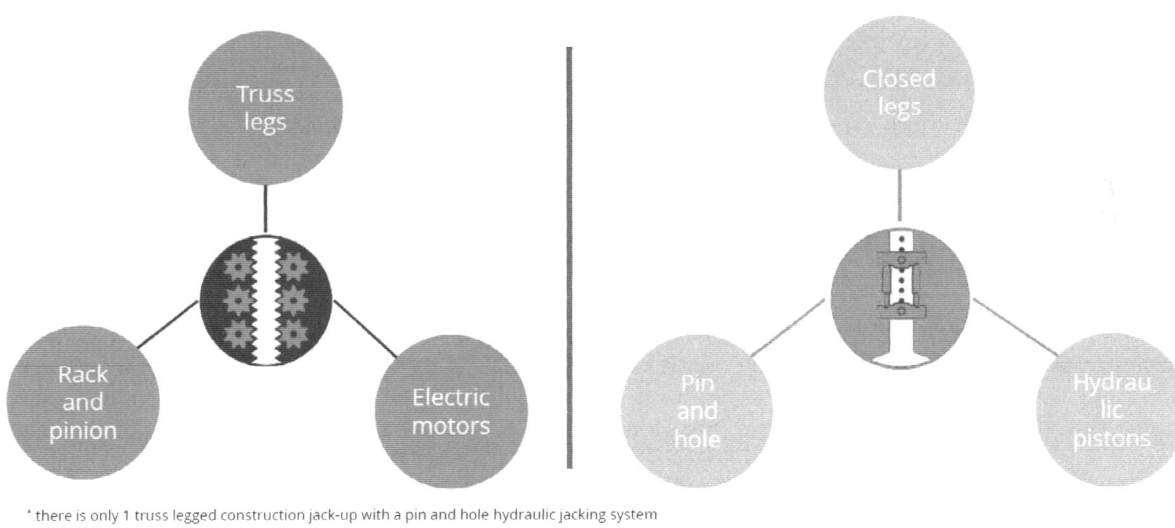

Figure 5: Truss & closed leg types form two distinct groups

Table 1: Representative selection of wind turbine installation jack-ups

	Unit 1	Unit 2	Unit 3	Unit 4	Unit 5
Number of legs	6	4	4	4	4
Leg type	closed or truss	closed	truss	closed	truss
Jacksys type		pin-hole/hyd	rack-pinion/elec	pin-hole/hyd	rack-pinion/elec
Preload	7500 t	2500 t	2500 t	9000 t	14000 t
Crane WLL	1000 t	600 t	600 t	800 t	2500 t
Crane type	pedestal	pedestal	pedestal	leg encircling	leg encircling

It is important to consider that different solutions may have different failure mechanisms and therefore different risk profiles. Without diving into details, at an identical site, the consequences of significant settlement are different for a four legged jack-up with truss legs and a leg encircling crane than for a six legged unit with closed legs and a pedestal crane.

The next consideration is the soil conditions. For some soil profiles exceeding the achieved preload has no discernible consequences, as would be the case for very dense sands. For other foundations the operator could notice settlement and the resulting increase in structural loading, but this remains a manageable problem as long as there is sufficient structural strength and increase in bearing capacity of the soil. Finally, there are sites where additional settlement could have catastrophic consequences, for example locations where punch through is expected or where the bearing capacity does not increase sufficiently with penetration to counter the additional loads due to platform inclination, which leads to increased crane off-lead and hence larger crane induced loading.

In terms of risk, let's categorize three levels for a case matrix that considers both jack-up as well as foundation properties.

- Low risk — failure has no impact on operations
- Medium risk — failure results in a temporary stop, corrections are required
- High risk — failure results in loss of life and loss of platform

To each of these risk levels, a probability of failure may be set by the stakeholders, weighing the consequences of accidents. In the low risk case, exceedance of the preload is considered acceptable, so foundation stability does not impose a real limit. For the high risk case foundation stability imposes an important limit. Consider 1/10,000 as an acceptable probability of failure for a single event, just as an example.

With the probability of failure for the extreme event now set, an important consideration is the probability of failure over the lifetime of the jack-up. Wind turbine installation jack-ups operate near the boundary of failure practically on a daily basis. And it can intuitively be understood that the likelihood of ending up on the wrong side of the safety threshold increases with the number of extreme events experienced over a lifetime. You have to be on the safe side every single time. Statistically the probability of failure over the lifetime is:

$$P_{f\ lifetime} = 1 - \left(1 - P_{f\ event}\right)^n$$

Where n is the number of events during a lifetime.

Assuming the jack-up performs one high risk lift per week over a lifetime of 25 years, that means the jack-up would experience 1 x 52 x 25 = 1,300 high risk events in its lifetime. If the probability of failure for each of those events is

1/10,000, then the probability of failure over the lifetime P_f, would equal 0.122. So 12% of the jack-ups would experience failure over the lifetime.

Note that the example above uses arbitrarily chosen values intended to make the point that the repetitive nature of wind turbine installation has to be taken into account. This example portrays the factors considered in safely optimizing performance of these jack-ups, and is not a recommendation of acceptable probability. Acceptable probabilities of failure should be agreed upon by the industry stakeholders considering the risks.

Limit state functions and stochastic variables

In the limit state function for crane operations with a wind turbine installation jack-up the resistance is the preload and the load is the load at the foundation, so the expression of the margin can be written as follows:

$$margin = preload - load$$

Where both the applied load and the resistance are governed by weight related loads. A list of the components to account for in operating conditions are:

- Weight and cog offsets
- Friction
- Environmental loads (Wind, wave, current)
- Inertial loads

The weight loads during operations account for 80% to 100% of the total load on the foundation. Depending on the quality of the measurement instrumentation (accuracy, repeatability, reliability, or possibility to detect faulty sensors), the procedures and the effort gone through, the weight loads can be known with high accuracy and precision.

If loads are measured in the jacking system, the friction loads between legs and guides are typically the largest source of uncertainty. Friction loads introduce a bias in the measurement, which reduces accuracy. The direction of the bias depends on the movement of the leg relative to the guide structure. The magnitude of the load depends on many factors, e.g. whether greasing is applied, as built tolerances, jacking system design, site conditions that induce leg splay, etc. In most cases friction loads are below 10% of the total load of the leg (based on total friction load divided by number of legs, expressed as fraction of the actual individual static leg load level), but for a site that induces leg splay the friction may well exceed that value. Although the average of around or below 10% may seem small, individual variation per leg may be well beyond 10% and poses an additional challenge.

During crane operations the environmental and inertial load are typically small, somewhere between 0% to 10% of the load on the foundation. During preloading, the effect of environmental loading is normally even lower. Although the relative magnitude of the environmental load contribution is small, it a larger source of uncertainty.

Statistical modeling is needed to assess the expected maximum environmental load during the operations. Referring to the different leg types as indicated in

Table 1 (truss type versus closed type), the load levels may be assessed differently per leg type. The reasoning behind this is that the relative importance of drag loads versus inertia loads (quadratic versus linear dependency on wave height) leads to a different extrapolation to for example most probable maximum loads.

Resistance: Preload

The objective of the preload operation is to establish sufficient foundation capacity to resist leg loading during operational and storm conditions. For a wind turbine installation jack-up, preloading is typically performed by carrying the hull weight on pairs of diagonally opposed legs in turn. This can be done actively by pushing the legs (limited by jacking capacity), or passively by reducing the loads on one diagonal and therefore increasing the loads on the other pair (limited by the static holding capacity), as shown in Figure 6.

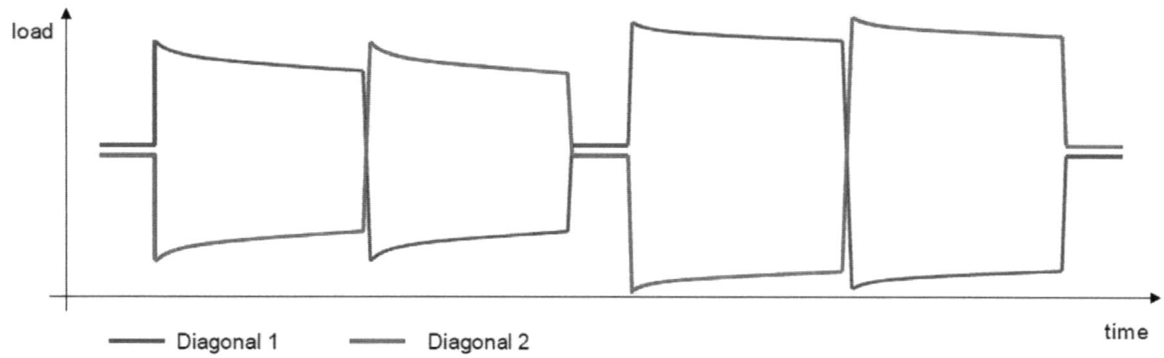

Figure 6: Typical example of diagonal preloading.

Preloading can be done fully, utilizing all the elevated weight available, or it can be optimized based on the planned operations or applicable storm conditions. The target preload level is determined based on a site specific assessment covering the governing conditions. A distinction could also be made between the achieved preload capacity for operational and survival conditions as illustrated in Figure 7. This distinction in resistance stems from the difference in the duration in which these loads are present. Survival loads are dominated by environmental loads and have a short duration, the order of magnitude is seconds. Crane operational loads are weight dominated and have a long duration, the order of magnitude is minutes to hours.

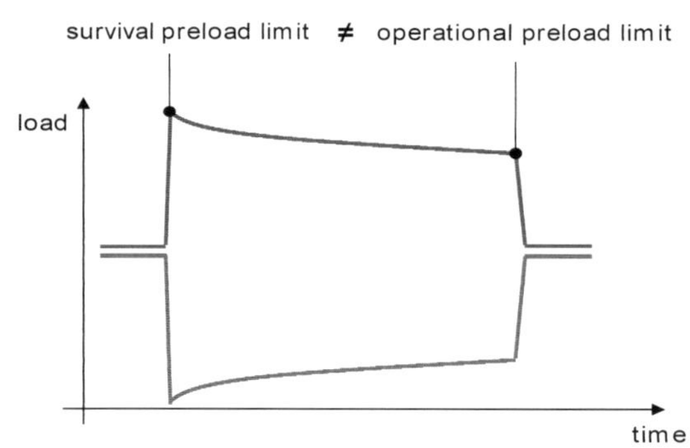

Figure 7: Proposal for distinction between survival and operational preload

Figure 8 shows jacking system load data from a real preloading operation. The values are nondimensionalized by plotting the ratio of the measured leg load over the maximum preload for that jack-up.

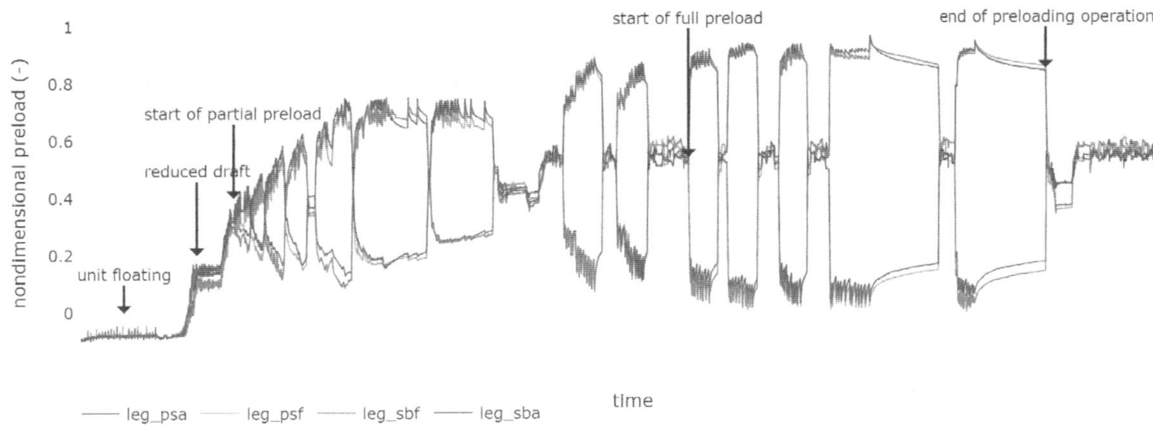

Figure 8: Typical preloading procedure

Starting at the beginning of the time trace, the jack-up is afloat. The load on the legs are negative since the legs are hanging in the jacking system. After a while the legs touch bottom and as the jacking system lifts the hull out of the water the loads on the legs increase. First the foundation is preloaded partially. Here the hull is still partly in the water and affected by the waves. The time varying load component is clearly visible.

After partially preloading the operator is sure the foundation is able to take on the full weight. The hull is lifted completely out of the water, just clear of the waves. Now the full elevated weight is available for preloading. In the example of Figure 8, the operator is faced with challenging soil conditions, such that a stable preload level is not readily achievable. As soon as the operator increases the load on the foundation, the load reduces due to settlement resulting in the sawtooth pattern. Finally, the operator calls it quits as the settlement seems to stabilize at about 85% of the maximum jacking system preload.

From the above time trace, is there a clearly defined achieved preload? Currently this is tough to answer as a practical definition of the achieved preload does not exist [1][2]. For clarity, the following definitions of preload are made in this paper:

- Maximum jacking system preload — Maximum preload that can be achieved by the jacking system
- Maximum achievable preload — Maximum preload that can be achieved on basis of the elevated weight, offset of center of gravity with respect to center of legs and quasi-static environmental load contributions. Note that the maximum achievable preload values typically vary per leg.
- Required preload — Preload level as assumed for the Site Specific Assessments (SSA's) for both survival and operational condition.
- Achieved preload — Level of preload achieved during the in-situ preloading process, according to predefined criteria determined by stakeholders.

Keep in mind that the achieved preload established during these operations imposes a very real limit. Ideally this achieved preload should be equal to or above the required preload. Also, this limit needs to be defined clearly to start with rationally quantifying the reliability. Due to this gap in guidance the criteria used to evaluate the achieved preload differs across operators and currently industry practice is to quantify the achieved preload using human judgement. Getting the stakeholders to agree on the algorithms and criteria used to determine the achieved preload and store the result in a digital database would not only be practical, it would be a leap in safety.

Load: hull weight and crane loads

A schematic version of a lifting operation was shown in Figure 1 and 3. Figure 9 and Figure 10 below show the same concept, but is based on real data from the crane and jacking system of a wind turbine installation jack-up in operation.

The operations took place after the preloading operations from Figure 8. The working load of the crane and leg loads are nondimensionalized. The nondimensional working load of the crane is the working load divided by the maximum working load (right axis of Figure 9). The nondimensional leg load is the leg load divided by the maximum jacking system preload (left axis of Figure 10).

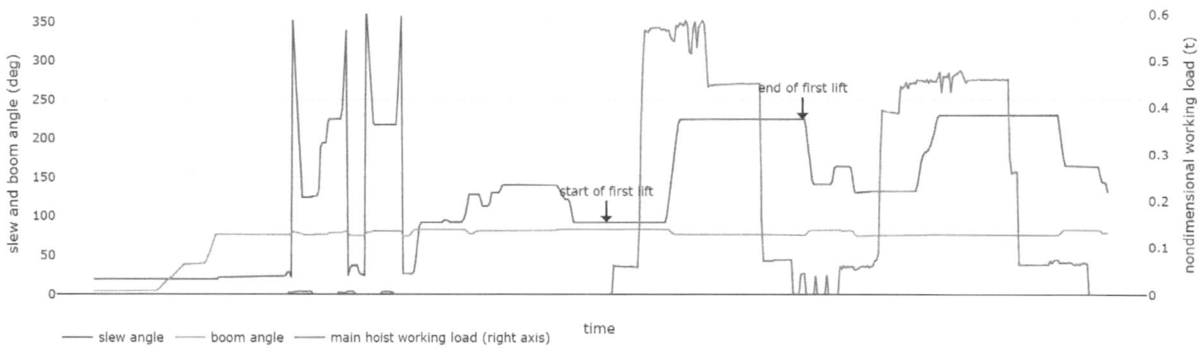

Figure 9: Typical crane operation

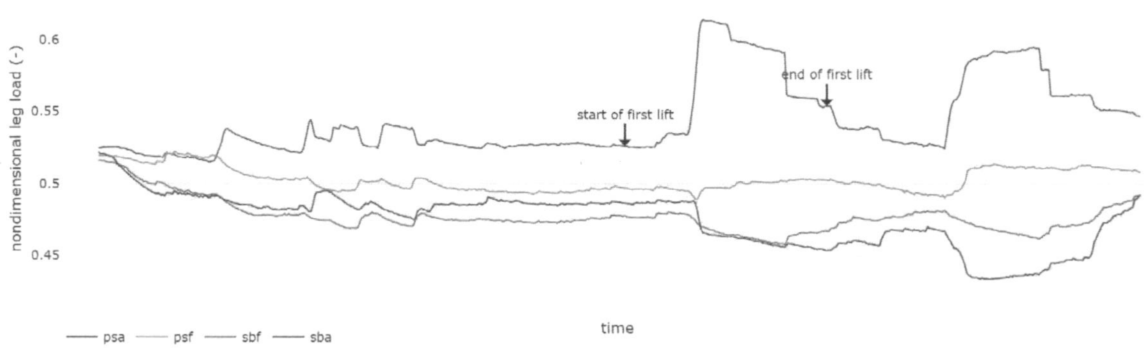

Figure 10: Typical foundation load during crane operation

These time traces illustrate how the crane operations influence the leg loads. At the start of the time trace, the crane operator lifts the boom, then performs some small lifts with the main hoist. Finally, a first piece of payload is connected and lifted into place. Then the crane performs a second lift and the time traces end.

In the example the effect of the crane operations is clearly visible, but in magnitude the effect is limited. There is also a large margin between the maximum load on the legs during operations, approx. 65% of the maximum jacking system preload, and the preload that was established, which amount to approx. 85% of the maximum jacking system preload as discussed in the previous section. Figure 11 brings the preload and load data from the example together schematically.

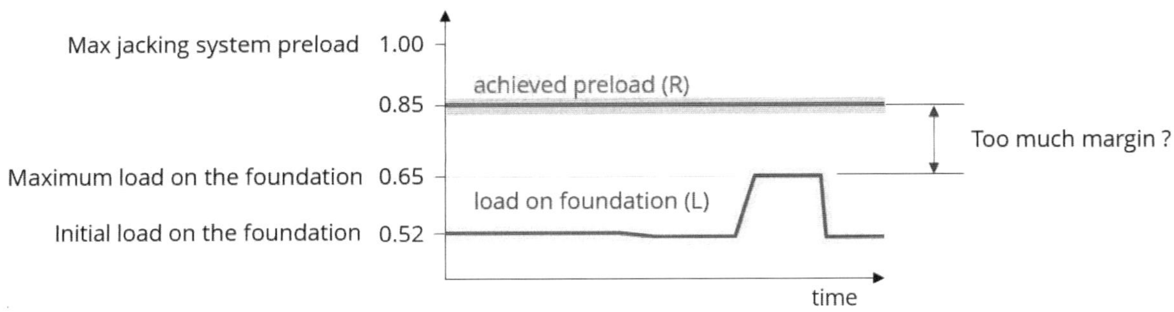

Figure 11 Achieved preload and load on foundation for typical case (values from example)

Considering the data this operation was undoubtedly safe, but perhaps it is was too safe. Depending on the point of view it can be concluded that the preload could have been reduced further, that there is excess preload capacity for this unit, or that the crane potential for this jack-up can be increased dramatically. Whatever the perspective taken, there is room for improving the design and the operation.

Before continuing on quantifying with the section on quantifying a rational margin, below some key takeaways from these sections on load and resistance:

- How critical a lift is in terms of foundation stability depends on the achieved preload and the load on the foundation
- The maximum achievable preload depends on the elevated weight, but the achieved preload is determined by the operators. For a four legged unit the achieved preload value should be greater than 25% of the elevated weight, but will most likely be lower than 50% of the elevated weight
- The initial load on the foundation during crane operations depends on the elevated weight, but the magnitude of this load is determined by the operators. For a four legged unit the initial load on the leg can be distributed such that a leg accounts for as low as 0% of the elevated weight, up to 50% of the elevated weight
- The maximum load on the foundation depends on the initial load determined by the operators, from there the the peak of the load is determined by the crane and the lift it is performing

In essence wind turbine installation design is about combining the right crane with the right jack-up. It is possible to outfit this jack-up with a much larger crane, a crane twice the current size, provided that an effort is made to accurately define the limit and ensuring that that limit is respected. The typical example of a wind turbine installation jack-up, is the jack-up not yet reaching its full potential.

Safety factor

Having addressed the target reliabilities, and the limit state function and its variables, it is time to quantify the margin with safety factors.

The trivial case from earlier in the paper is the case labeled with low risk for operational conditions when the foundation capacity is exceeded. In this case the foundation safety can be assumed to be covered if it is demonstrated that no further relevant settlement occurs, as there is no consequence related to its exceedance. If there is no noticeable settlement, then there is no additional (relevant) load effect due to redistribution of loads.

For the high risk case, additional consideration is required. Using the basic structural reliability model, the safety factor required to attain the target probability of failure for an extreme event can be determined as a function of the standard deviation σ of the load and resistance. For example, consider a wind turbine installation jack-up with closed cylindrical legs and a maximum preload of 9,000 t, (unit type 4 from

Table 1). Assume that the target reliability is on average 1 failure every 10,000 events and that both preload and foundation load can be modeled with normal distributions and that their standard deviations are equal.

Table 2: Safety factor estimates for crane operations with a 9,000 t preload jack-up and Pf = 1^{E-4}

P_f	β	σ_L & σ_R	σ_M	safety factor
1E-4	3.72	200 t	$\sqrt{2}\cdot 200$ t	1.13
		100 t	$\sqrt{2}\cdot 100$ t	1.06
		50 t	$\sqrt{2}\cdot 50$ t	1.03

Table 3: Safety factor estimates for crane operations with a 9,000 t preload jack-up and Pf = 1^{E-7}

P_f	β	σ_L & σ_R	σ_M	safety factor
1E-7	5.20	200 t	$\sqrt{2}\cdot 200$ t	1.20
		100 t	$\sqrt{2}\cdot 100$ t	1.09
		50 t	$\sqrt{2}\cdot 50$ t	1.04

The standard deviations listed in Table 2 and Table 3 above are estimates, ball park figures since the work of modeling the stochastic variables has not yet been performed. The standard deviation of 200 t is approximated from time domain simulations modeling operational environmental and inertial loads representative for this particular unit, water depth, soil and wind, wave & current combination. The standard deviation of 100 t and 50 t are to analyze the sensitivity of the safety factor to the uncertainty.

For this example with three semi-arbitrary uncertainty levels, assuming normally distributed load and resistance levels, and zooming in on the relatively easy quantifiable environmental and inertial load contributions, the upper bound for the safety factors would ranges from 1.13 to 1.20 depending on the target probability of failure. Depending on further subdivision, and effort put in quantifying all relevant uncertainties, specific for the unit and application at hand, actual values will differ.

FURTHER CONSIDERATIONS TO INCREASE SAFETY AND PERFORMANCE

The safety factors that should be created with input from all stakeholders are a great tool, but it is not the end result. The goal is having performant, reliable and most importantly safe wind turbine installation jack-ups out there. Specific to crane operational conditions, this means not overutilizing the foundation capacity and improving the safety and performance of these units further. This section outlines additional considerations to potentially increase the safety and enhance the operation capacity of wind turbine installation jack-ups.

Human error

Looking at failure only from structural reliability point of view is forgetting a very important source of failure: human error.

It is impossible to know everything in advance, and not all cases can always be accounted for. Engineers use engineering judgement to generate conservative cases, but erroneous assumptions are sometimes made. Engineers, being human, can make mistakes like faulty assumptions.

Furthermore, consider that the governing component for load and resistance is not controlled by the weather, but by human operators. One little nudge on a joystick of the jacking system and the weight on the foundation increases; Don't push the joystick enough, and the target preload might be missed. Increase the outreach for the crane too much, and the foundation capacity may be overutilized.

For wind turbine installation jack-ups, the industry relies mainly on people and processes to keep things safe. Human judgement is relied upon to quantify the preload. Procedures later prevent the crane operator from exceeding limits because he sticks to his lift plan, but other than operator experience and procedures there is nothing that prevents the crane operator from exceeding the achieved preload. The crane system is unaware of the limit imposed by the jacking system and soil.

The probability of human error can be reduced by investing in people, processes and technology. Sound engineering and good procedures as well as well-trained and experienced operators working with reliable systems are needed to reach the wind turbine installation jack-ups full potential.

Instrumentation and data collection

Consider investing in instrumentation to reduce uncertainties, which would allow a decrease in the required margin at an identical safety level.

As illustrated in Table 2, if the level of uncertainty is reduced, less margin is needed to achieve the same level of safety. This simple principle is illustrated in Figure 12, where the effect of reducing the standard deviation from 200 t to 100 t on the required margin to reach the same probability of failure of 1 in 10,000 is shown. Accurate, precise and reliable instrumentation is important to reduce uncertainties and consequently enlarge the operational potential.

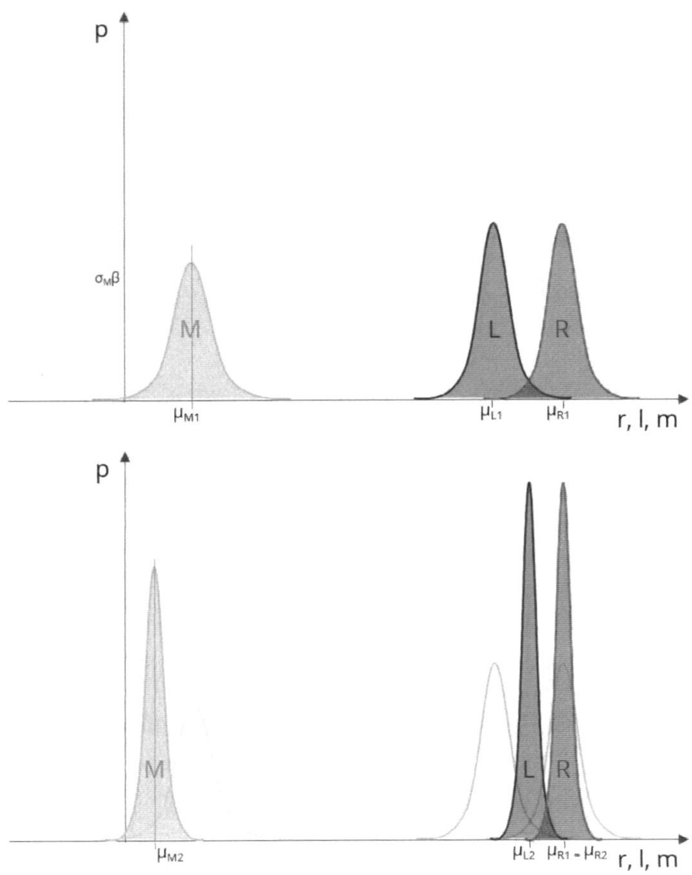

Figure 12: Decreasing uncertainty from σ = 200 t (top) to σ = 100 t (bottom) for 9,000 t preload (Table 2)

Optimizing the leg load distribution

Contrary to three-legged oil and gas jack-ups, four (or more) legged wind turbine installation jack-ups have a large extent of control over the loads on the foundation. Moving weight from one leg diagonal to another, maintaining static equilibrium, can be done fast and well controlled. That means that before the onset of a preload operation weight on the crane leg could be transferred to the other legs, effectively increasing the margin for the foundation of the crane leg and therefore increasing the allowable crane outreach and working load. This is the basis of pre-torque. It allows the operators of wind turbine installation vessels to use their equipment to its full potential. Once a reliable system is created, the design and the use of the wind turbine installation jack-up can be truly optimized.

LINK TO BIG DATA TECHNOLOGY TO MAKE IT POSSIBLE

The issues blocking the full potential of wind turbine installation jack-ups are clear. The basic principles behind the solutions are relatively simple. However, the technical implementation of these solutions is complex and remains to be proven.

The most complete way to optimize the safety and performance of a wind turbine installation jack-up, is to take advantage of the vast amount of data that can be gathered and translated into information in real time. Big data technology is a key component to making this possible.

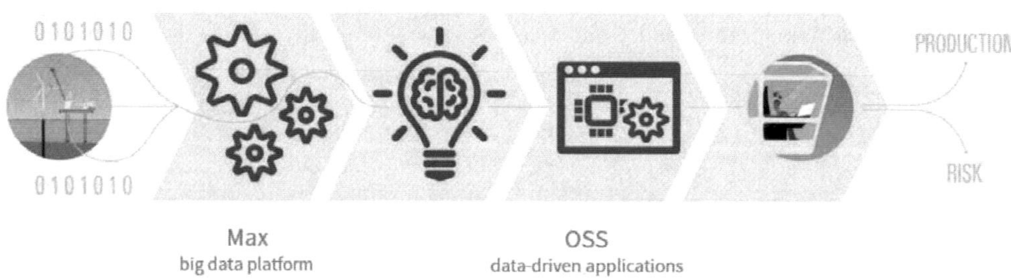

Figure 13: Flow of Data from equipment through operator application

Figure 13 outlines the process of collecting useful data, processing, storing and finally interpreting in an application. Through these steps, the data becomes useful information, captured in an application, accessible to the operators to assist them in making the proper judgement calls.

Data is needed from the different control systems like the jacking system and the crane which are by nature isolated environments. By combining the individual environments and adding additional instrumentation, the reliability and/or accuracy of the system can be increased. Before the information can be processed, the data is gathered and prepared on a communication platform and made suitable in downstream applications. Finally, the data reaches the end user through a data driven application that quantifies limits in real time, safeguarding them from unsafe operation, while increasing the equipment performance.

GustoMSC relies on the Industrial Internet of Things platform Max™, conceived by NOV in 2015, to harness rig data. Max is not only a data delivery structure; it is also a complete development environment for creating data driven applications for edge and cloud. In this particular case, the focus is on the edge of the network, so near the source of the data. The Operator Support System (OSS) is a series of GustoMSC applications operating on the Max platform. The first two applications are OSS.Jack and OSS.Lift. OSS.Jack is the application that aims at setting a reliable achieved preload level, while OSS.Lift uses that preload as input to safeguard the end user during crane operations.

OSS.JACK

In its simplest form, OSS.Jack deduces the preload solely on jacking system data. In more elaborate forms additional instrumentation is required to enhance the precision of the measurements (e.g. additional independent/second leg load measurement system). It is this process of increasing the precision that allows operators to increase the reliability of their operations and/or the overall potential of their jack-up.

In its basic form, the achieved preload for operations can be deduced by monitoring the jacking system data. On basis of intuitive considerations, an achieved preload value can then be determined using two criteria:

1. Duration of the loads in the jacking system
2. Rate of change of the loads in the jacking system

Both criteria need an input parameter (time and rate of change respectively) which depends on the soil conditions (location) where the jack-up is performing the preloading operation.

It is the responsibility of the end user of the jack-up to specify these input parameters before the start of a preloading operation. Ideally, these parameters have been agreed upon between all stakeholders (e.g. warranty surveyor, operator, etc.).

When OSS.Jack has determined the achieved preload based on the two criteria (load duration and rate of change), it notifies any dependencies of the result (e.g. the edge device in the crane) and stores the values in the database. Event triggers automatically keep track of where the operator is in the jacking process. Additionally, custom client procedures as a series of steps that the operator follows can be programmed in and checked for consistency with other systems, or with criteria for safety and reliability.

OSS.Jack provides feedback to the jacking master through a graphical user interface. Algorithm input parameters related to preload acceptance criteria are provided through the OSS.Jack user interface. The basic output of the user interface is illustrated below in Figure 14. The interface gives insight as to what has happened recently, what is going on now and what is likely to happen in the near future (central time traces). It visualizes achieved preload levels (red bar on the left) and shows the predicted preload (blue bar on the right).

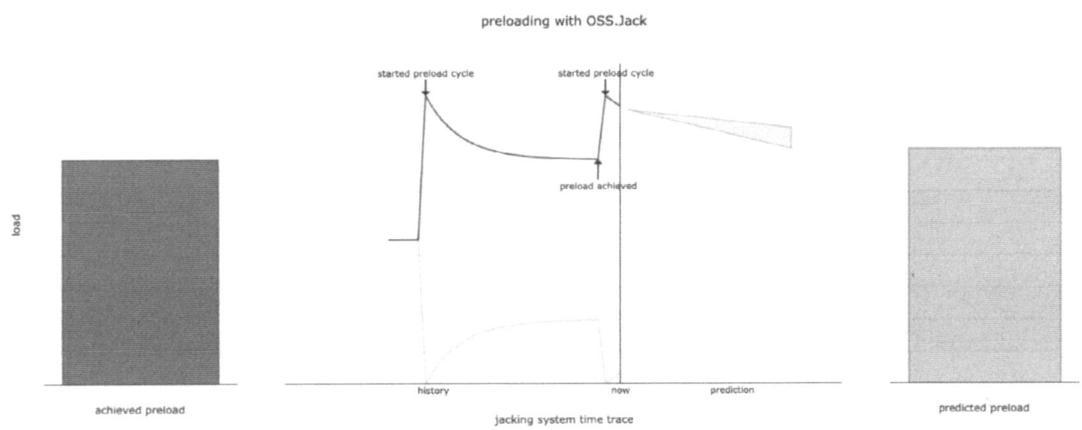

Figure 14: Graphic interface of achieved preload in OSS.Jack

An important improvement to preloading procedure is the addition of automated reporting of the performed preload operation and the achieved values is generated. This report is filed and can be shared with assigned parties as per client's preference. The report can be customized to the requirements of the client within the limits of the system. By automating this functionality, the user is less prone to mistakes in the transfer of information between interested parties. Furthermore, an approach on basis of agreed criteria can be followed consistently. Additionally, this information can also be provided as input directly to the crane operator through OSS.Lift.

OSS.Lift

To enable safer and more efficient lifting operations, GustoMSC offers OSS.Lift. In its basic form this application uses data from the vessel management system, crane and jacking system and processes it into information that increases the operator's awareness of the operational limits. Optionally a client specific procedure may be configured around the core of the application.

The basic version of the application uses a mathematical model of the jack-up and crane to assess the acceptability of the operation by evaluating the maximum load on the foundation accounting for the jack-up's stiffness, weight, environmental loads and guide friction in combination with the measurements and to compare that with the achieved preload. The combination of the model and real-time data is used to derive the real-time limits. The application has 3 user interfaces that can be used across all interested parties, for example: project engineers or crew, jacking system operator and crane operator.

Project engineer or crew on board

OSS.Lift provides the engineering tools required to verify what minimum preload is required to perform the oncoming lifting operation safely, effectively setting the target value for the preloading operations or verifying the target preload level.

By changing the relevant input variables such as preload, crane load, outreach and slewing angle, the user can quantify the required preload and load distribution to safely complete the lifting operation.

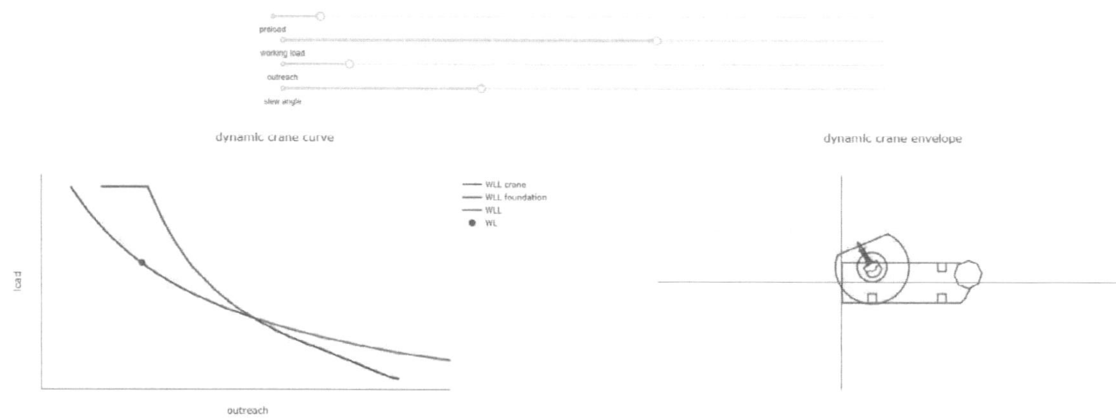

Figure 15: Graphic interface of crane lift planning OSS.Lift

Jacking system operator

The interface for the jacking system operator gives a clear overview of the loads on the legs, the achieved preloads and the utilization of jack-up's foundation. The interface clearly also shows the load components and the safety margin applied.

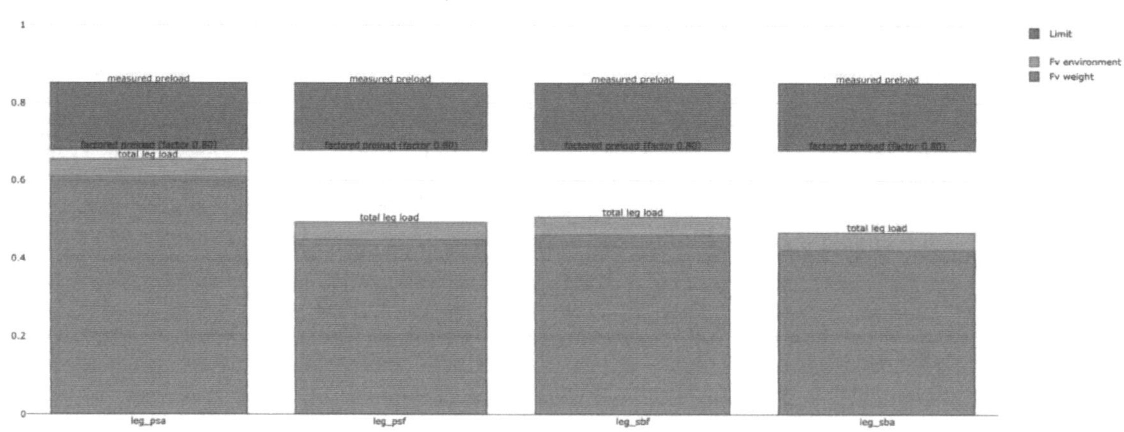

Figure 16: Graphic interface of jacking load monitoring OSS.Lift

Crane Operator

The interface for the crane operator gives a clear overview of the crane's real-time working load limit and envelope. This is a leap in safety, as now the limit imposed by the jack-up's foundations is considered within the crane system.

If two-way communication is enabled with the crane, the foundation limit can be accounted for within the crane's control and safeguarding system. In that case the crane system can prevent the operator from performing an unsafe operation, like maneuvering a lifted load outside its safe working area outlined by the crane envelope.

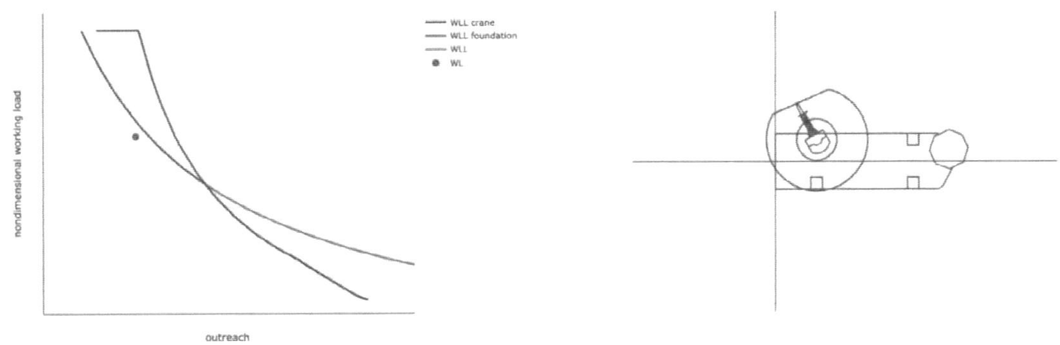

Figure 17: Graphic interface of crane operating envelope OSS.Lift

SUMMARY

The level of safety and performance of wind turbine installation jack-ups is currently suboptimal because clearly defined limits and target reliabilities for critical operations do not yet exist. This paper addresses the gaps that, from the perspective of the designer, prevent optimal use of wind turbine installation jack-ups and how this could be addressed:

- Target probabilities of failure considering the jack-up design, foundation conditions and resulting risks have not yet been determined by industry stakeholders
- Limit state functions, the mathematical descriptions of failure, have not yet been defined
- Preload, the essential variable that outlines the capacity of the foundation, has not yet been defined
- Data that can be used as input for stochastic models has not been processed
- Safety factors have not yet been determined
- Human judgement and procedures are heavily relied upon to prevent accidents
- Technology is insufficiently leveraged to improve safety and performance

To fill these gaps and unlock the full potential of wind turbine installation jack-ups, all stakeholders need to contribute: operators, regulatory bodies, designers, shipyards, equipment manufacturers. All have valuable insight to contribute to safe jack-up operation.

GustoMSC has proposed a suite of tools, OSS.Jack and OSS.Lift, that will give more useful information to engineers and operators that allow them to extend the capacity of their units. OSS.Jack monitors load duration and rate of change that allows the operator to calculate the settled foundation capacity as the achieved preload. OSS.Lift takes information from OSS.Jack and extends to useful information for crane operation, different interfaces allow this tool to align office engineers, jacking system operators and crane operators. The combination of OSS.Jack and OSS.Lift provides data driven insight needed to safely optimize the crane operations of a wind turbine installation jack-up.

ACKNOWLEDGMENT

With this paper our aim was to create reference literature how data-driven applications can make a difference for wind turbine installation jack-ups, and highlight some key issues to be tackled before we get there.

Special thanks to Andries Hofman and Thomas Lerchenmuller, for their insight and knowledge of wind turbine installation jack-ups, their willingness to share ideas and their time spent contributing to and reviewing this paper. For Frederik Vanhee, the co-creator and my partner in building the foundations of the data driven operator support system supported by Max.

I would also like to express the enormous amount of respect and gratitude I feel for NOV's Max Digital Services team. The amazing big data platform you are building really opens up this technology for a wider group of people. Without your mind boggling efforts, reaping tangible value from big data technology still seemed so far away.

REFERENCES

[1] https://gwec.net/gwec-market-intelligence-releases-global-offshore-wind-turbine-installation-database/

[2] SNAME, Technical & Research Bulletin 5-5A, Guidelines for Site Specific Assessment of Mobile Jack-Up Units, August 2008 with corrections from 30 June 2015

[3] ISO, 19905-1, Petroleum and natural gas industries – Site specific assessment of mobile offshore units 2016-01-15

SNAME
THE INTERNATIONAL COMMUNITY
FOR MARITIME AND OCEAN PROFESSIONALS

DESIGN OF A FREE RUNNING HIGH SPEED CRAFT MODEL FOR TESTING OFF THE COAST OF GALVESTON

Ahmed A. Hamada
Texas A&M University
Omar Khaled Sallam
Texas A&M University
Colton R. Bonnaure
Texas A&M University
Patricia Itzel Rodriguez
Texas A&M University
Jack Stason
Texas A&M University
Mirjam Fürth
Texas A&M University

Proceedings of the 26th Offshore Symposium, April 6th **&** 7th***, 2021, Houston, Texas***
Texas Section of the Society of Naval Architects and Marine Engineers

ABSTRACT

High speed planing craft (HSC) are used in a wide variety of marine applications, such as search and rescue, defense and surveillance, recreation, and as support and supply vessels to the offshore industry. Their design, improvement, and optimization mainly depend on a deep understanding of the dynamic interaction between the vehicle and the surrounding environment. Traditionally, measuring the dynamic behavior and response of the HSC due to waves has been performed in towing tanks, where measurements are recorded under ideal conditions. To get a more realistic idea of the performance of HSC, we aim to perform free-running tests in the Gulf of Mexico at the Texas A&M Galveston campus. The aim of the free running tests is to measure the behavior of the model under real-life conditions, such as realistic incident waves, wind, current speeds, and variable water depth regions. Designing such an experiment far from lab facilities is especially challenging. In this paper, a detailed step-by-step HSC experiment design workflow is presented to tackle the open sea testing challenges. The paper presents a design of a free running model to be built during the spring of 2021, which includes the basic geometric parameters selection, stability study, drag force prediction, required propulsion system's power estimation, manufacturing techniques, and sensors' data acquisition system design.

Keywords: High Speed Craft, scaled Model Testing, Free-Running, In-Site Testing, Preliminary Design

Introduction

During past decades, High-Speed Crafts' (HSC) use and demand has grown in multiple market categories including but not limited to; military, search and rescue and the recreational sector. The most common experimental method to investigate the performance of HSC is towing tank testing, which represents ideal surrounding environment with limited wind effects, variation in current speeds, water depth and constant incident wave direction angle.

Free-running tests are performed in open-sea where more realistic and random conditions take place (von Ellenrieder 2013). However, they are more complicated to do, specifically due to the same features that make them more realistic, cross flow and wind, varying sea states and water depth, and a model that is free to move in all degrees of freedom. Furthermore, the model need to be self-propelled and fitted with a Data Acquisition System (DAQ). Due to the difficulty of open-sea free-running tests, they are becoming increasingly popular do in numerical simulation, to provide an increased controllability (Shen et al. 2014; Gaggero et al. 2017). Self propelled models have long been used in maneuvering basins, providing some of the benefits of a free running model while allowing for total control of the environment (Lewandowski 1995). However, these types of basins are becoming rarer and are more expensive to operate than conventional towing tanks.

Grenestedt et al. (2015) introduced a high-speed USV, called Lehigh Ocean Research Craft Autonomous (LORCA). LORCA was Compact with a hull length of 1.2 m, weight of 5 kg, rigid one-piece carbon fiber model, and self-righting. The LORCA's team and other research groups, such as in Breivik et al. (2008), Valada et al. (2014), focused on the trajectory following, straight-line target tracking, and multi-robot autonomous control, respectively. Nevertheless, they did not present a detailed design workflow of their HSC.

Jiao et al. (2018) conducted a comparative study of sea-keeping for a Chinese navy ship with a hybrid monohull of 134 m length, using a small-scale hull in the towing tank of Harbin Engineering University and a large-scale free-running model in the Bohai Sea in China. They showed that the free-running in the open sea is more realistic than the towing tank test as the free-running test provided close vertical responses to that of the actual ship, unlike the towing tank test, which overestimated the dynamic responses.

Recently, a naval and ocean technology laboratory at Texas A&M University, called FurthLab, introduced the first step to perform free-running tests for HSC by outlining the preliminary design and the testing plan of an open-sea free-running model Wang et al. (2020).

This research aims to continue this work by presenting a detailed design sequence for a free-running planing model, to be tested in the open sea. This paper details the current steps in the design cycle: such as stability, prediction of the drag force and the required propulsion system's power, manufacturing techniques, control, and sensor selection. This paper concludes with a description of the further work needed to build the model.

Model Specifications

Proposed Usage Scenarios

The initial test of the designed free-running model will focus on running position and resistance in calm water and acceleration in waves. Further, tests will focus on studying the flow detachment using spray detaching elements, such as spray rails (Seo et al. 2016; Lakatos et al. 2019; Clement 1964) and spray deflectors (Molchanov et al. 2019; Olin et al. 2017; Wielgosz et al. 2018). Thus, the designed HSC will be operated at various speeds and in different sea states. Furthermore, the tests will be performed during the daylight about 20 nautical miles from Texas A&M University Galveston Campus.

Design Constraints

The designed model must be compact and lightweight to make the model easy to handle and prepare by a few people. Further, performing a free-running test requires the model to be strong, and spacious inside to carry the hardware and sensors. The inside must also be shielded from water to protect the electronic components and sensors. In addition, the length of the hull was selected to be slightly larger than $6ft$ to follow the 28th ITTC (2014) recommendations for a free-running model and to be transportable with a standard truck vehicle. Further, a smaller length, L, could introduce scaling challenges. On the other hand, a larger model would be more cumbersome to manufacture and transport. For simplicity and easy comparison with exciting test data from a wide range of towing tank model tests, the hull will be V-shaped with a deadrise angle of 20° transom stern, hard chines, and flat buttock lines. Further, the Speed-Length Ratio, defined as in equation 1, must be larger than 3 to reach planning (Savitsky 2003). Thus, the cruise speed must be at least $7.35knots$ at model scale.

$$SLR = \frac{V_k}{\sqrt{L}} \tag{1}$$

where V_k and L are the speed in $knots$ and length in ft of HSC, respectively.

Design Workflow

The design of the HSC model follows an iterative design sprial that revolves around the model's dimensions, propulsion, weight, stability, and structure, as shown in Figure 1. The design sequence starts with

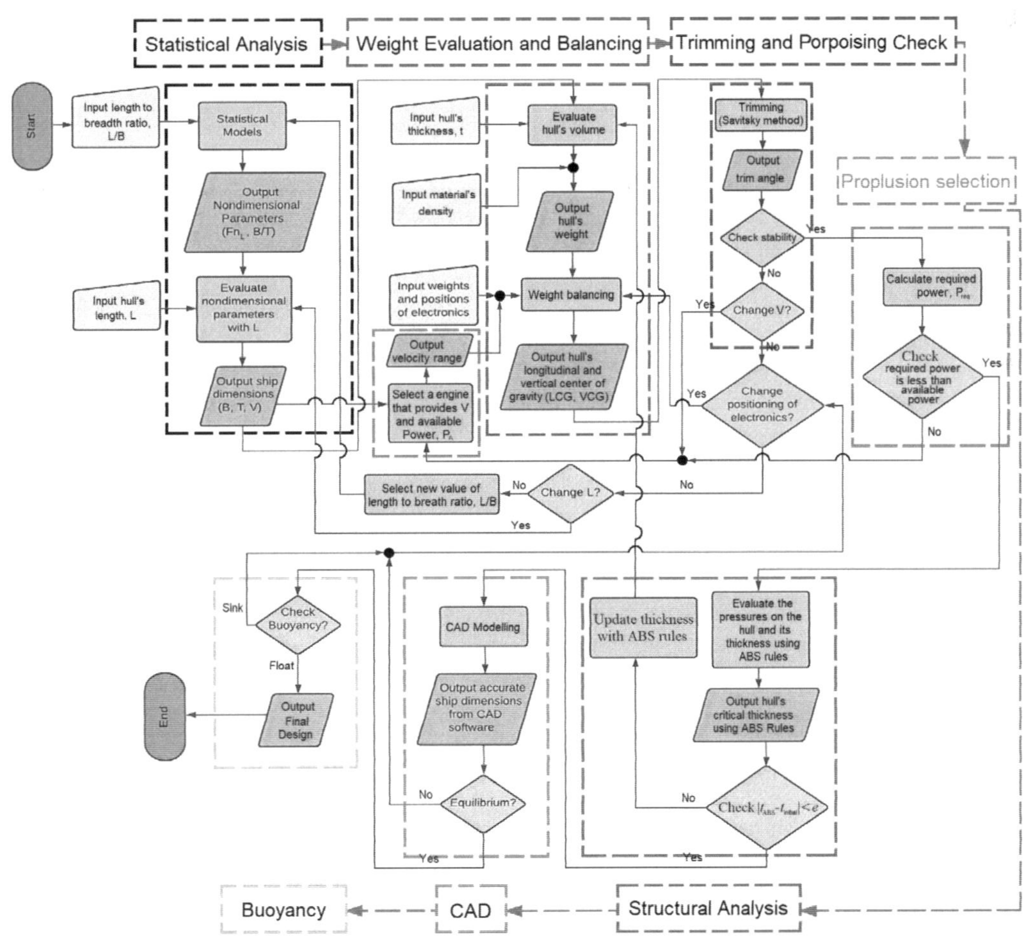

Figure 1: Design workflow for free-running model of HSC.

a parametric analysis, by selecting a length to breadth ratio, L/B, from the gathered data of existing HSC. Then, Fn_L and B/T are determined. By choosing an overall length, the breadth, B, draught, T, and velocity, V, are calculated from the statistics. Following this, a motor can be selected, that achieves a reasonable speed range based on the statistical analysis.

Furthermore, the main particulars are then evaluated to estimate the weight of the hull using the material density and hull thickness. Then, the electronics are incorporated to estimate both the total weight, the distribution of that weight and the Longitudinal Center of Gravity (LCG). These estimations are then used as to calculate the equilibrium trimming angle and check the longitudinal stability of the model at running speeds, using the Savitsky (1964) method.

If no equilibrium is found, this can be addressed by: adjusting the LCG or total weight, a new length can be selected which will alter the dimensions of the model based on the statistics, or the same length is kept and a new length to breadth ratio is chosen.

If a stable running trim is obtained, the required power is checked against the available power, provided by the motor, to ensure the selection of the motor is correct. If not enough power is provided, a new engine is selected and the procedure is repeated. When the required power does not exceed the available power of the motor, the model will be drafted in 3D CAD software, where accurate values for weight, center of gravity, etc. can be obtained. These new values are, once again, utilized in the Savitsky (1964) Method to recheck that a reasonable running trim can be achieved. If the model is determined to be porpoising stable over a wide range of velocities, provided by the motor, a final check for static buoyancy is performed. The model can be considered for a detailed design and subsequently built if it has successfully made its way through this design process.

Platform Design

Statistical Analysis

A parametric study is performed to ensure that the design model is representative of HSC in general since no specific parent hull has been chosen. For this purpose full scale HSC with public available data in three categories (search and rescue, coastal guard, navy) have been used. The parametric analysis is used to give an indication of the variation of the design parameters within each category. The Froude number, Fn_L, the Length to Breadth ratio, L/B, the Breadth to Draught ratio, B/T, and $M/(LBT)$, are used in the statistical analysis, see Figures 2, 3, and 4. Further, these figures provides possible combinations and limitations of the design parameters for the free-running HSC in an open environment. Thus, the results from the statistical analysis was used as a start point in the design procedure.

Material Selection

The materials being considered for the composition of the hull are: carbon fiber, fiberglass, and foam. Each of these materials have their own unique characteristics that fit the necessary requirements for a functioning HSC that will enable the model to be lightweight in order to achieve fast speeds and be able to withstand the force of slamming when the craft planes in waves.

Carbon fiber is one of the strongest, lightest materials available for a multitude of applications from golf clubs to air crafts, race-cars, and most importantly ships (Greene 1999). High tensile strength means carbon fiber is significantly stronger than steel or aluminum thanks and at the same time it manages to be exceptionally durable with a high strength to weight ratio (Khan et al. 2011). It's strength significantly out preforms the other two materials when the proper epoxy resin is applied.

Fiberglass shares many of the same characteristics as carbon fiber with both of them being recognized as strong versatile materials that are lightweight, corrosion resistant, and low maintenance (Blount 1982). Fiberglass differs from carbon fiber in that it is heavier, has a lower tensile strength, and it is made out of

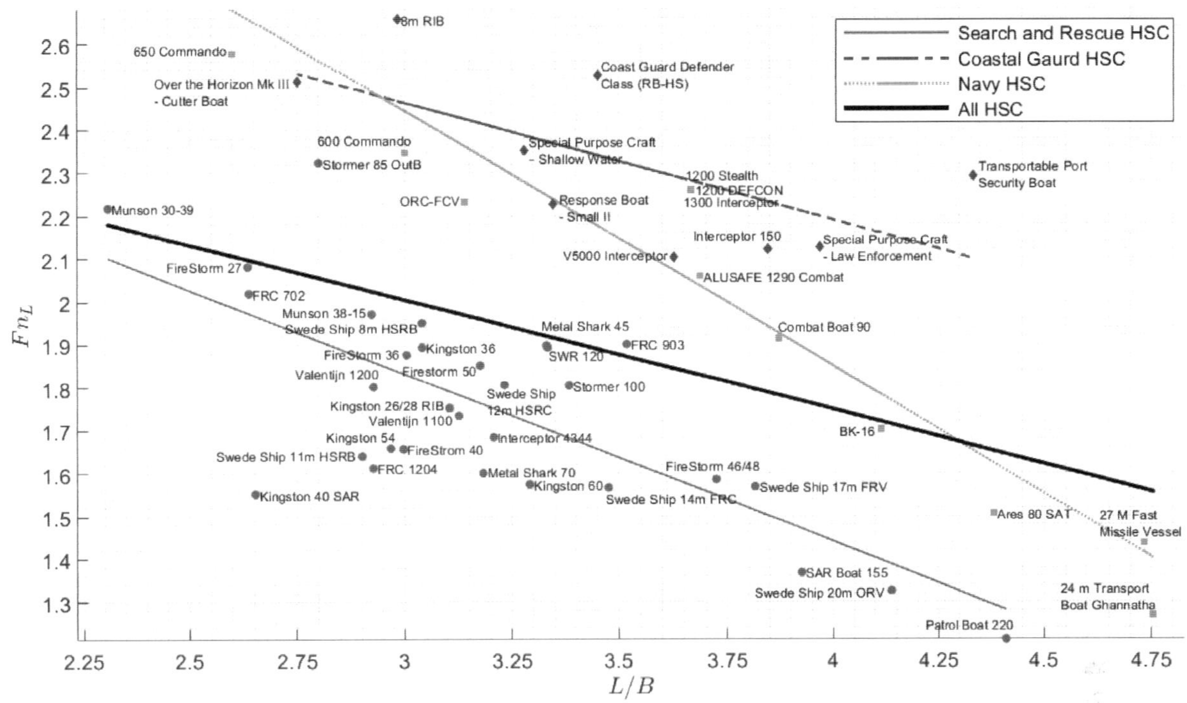

Figure 2: Variation of Froude number with Length to Breadth ratio for different categories of full-scale HSC

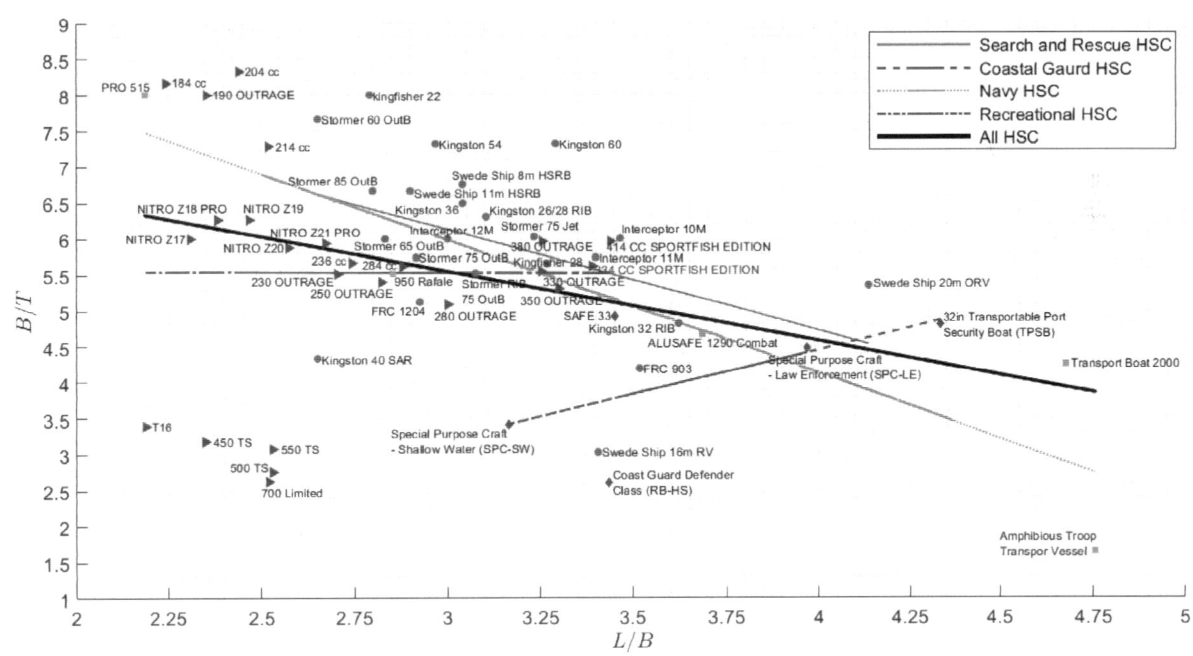

Figure 3: Variation of Breadth to Draught ratio with Length to Breadth ratio for different categories of full-scale HSC

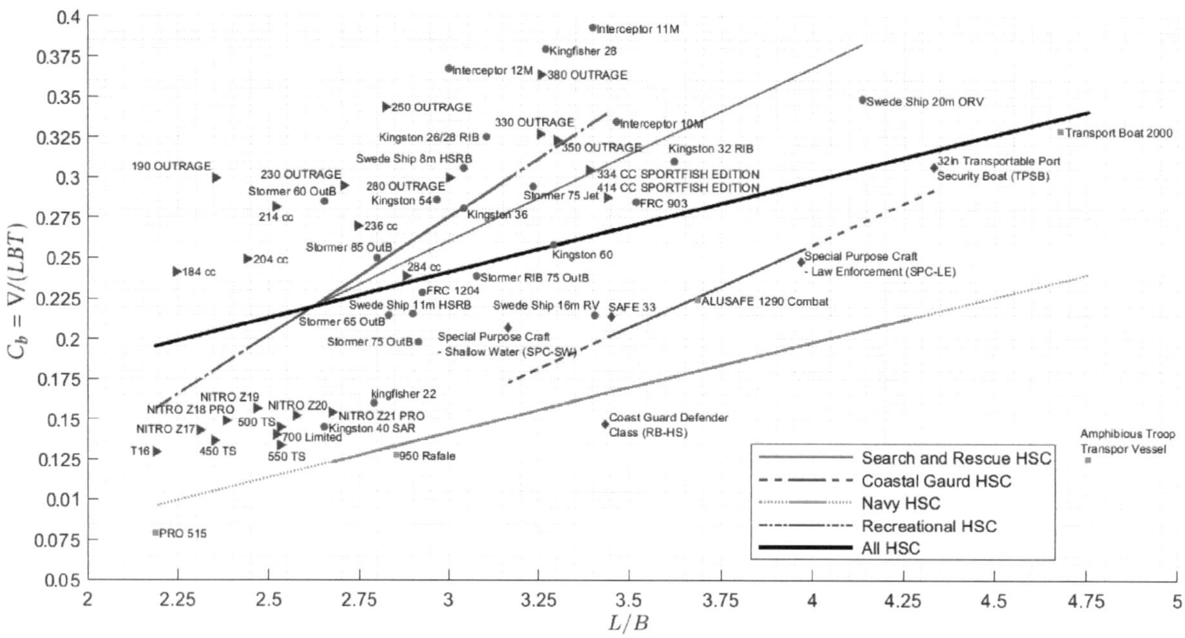

Figure 4: Variation of weight to Length-Breadth-Draught ratio with Length to Breadth ratio for different categories of full-scale HSC

layers of glass fibers unlike the continuous block carbon fiber is constructed from. Also the most notable difference between the two is that fiberglass absorbs more vibration energy than carbon fiber (Murčinková et al. 2019). Keeping in mind that one of the main issues of a HSC deals with when it begins to plane across the waves, it experiences vibrations on the craft which dampen the energy and thus diminishes the overall speed of the craft, making fiberglass more applicable in this case. One of the drawbacks to constructing a carbon fiber ship is that the material is quite expensive to manufacture and is one of the main reasons why this material is not as popular as fiberglass Das (2011).

Zisimopoulos (2016) noted the appeal for fiberglass because of its widespread use for boats due to its low cost to manufacture while still managing to be durable, fire resistant, and weatherproof. Now, there are in general two types of fiberglass, S-glass and E-glass, the former is considered as the more viable option of the two since it has better fatigue resistance, higher tensile, impact, and flexural strength. On the other hand, E-glass is more commonly used in boating since it's not as expensive (Greene 1999). The two types have the same chemical composition, but the main difference between them is the individual fibers of S-glass are spun more finely and of higher purity that contributes to its notable qualities.

However, there are some disadvantages to fiberglass, especially for those who own boats made of this material. In terms of maintenance and potential hazards, fiberglass requires a new coat of gel every five years because the coating ensures that the small glass particles the material is made out of are not exposed. Exposure whether by touch or inhalation to glass particles can be hazardous to an individual (Cooper 2019).

Foam is the last of the three materials chosen for the hull manufacturing, because just like fiberglass and carbon fiber, it is lightweight, has thermal insulation, high energy absorption, and it has a durable structure that allows it to be stiff and waterproof. There are a multitude of different types of foam each with their own set of characteristics due to the way their molecular structure is either cross-linked or interlocked (Editors 2007). In addition to the thermo-mechanical characteristics of the foam, what makes it so unique is the ability to work in tandem with the other two materials in a composite sandwich structure where the hull has a foam core with fiberglass or carbon fiber laminating the surface (Editors 2007). Thereby, it creates a boat with exceptional mechanical properties and dimensions thanks to it marrying either the carbon fiber or fiberglass.

Weight Estimation Of The Hull

The initial estimate for the weight of the hull of the model is performed by splitting the hull into three portions: the transom, mid-ship, and bow, as shown in the left drawing of Figure 5. Using the hull thickness, material, and deadrise angle, the weight of each section can be estimated independently before calculating a total weight.

The transom and mid-ship sections weights can be calculated simply based upon the beam, thickness, length, depth, and deadrise. Figure 5 depicts the geometry of the model.

$$W_{transom} = \rho g t \left(BD - \frac{B^2}{4} \tan\beta \right) \tag{2}$$

$$W_{mid-ship} = 2L_{mid-ship} \rho g t \left[\frac{B}{2\cos\beta}(1-\sin\beta) + D \right] \tag{3}$$

$$L_{mid-ship} = L - L_{bow} - t \tag{4}$$

For the high-speed craft model the bow volume is assumed not to vary with overall length. The length of the bow portion will remain constant in design considering it only is essential before the boat reaches the planing speed. The depth of the model will remain constant and, therefore, not affect the bow volume. Using a parent bow construction, multiple bow sections are constructed in SolidWorks to develop a curve that represents bow volume dependent on the beam of the craft. With the bow volume to beam relationship, for a given thickness, and material density, the weight of the bow section can be calculated.

$$W_{bow} = \rho g V_{bow} \tag{5}$$

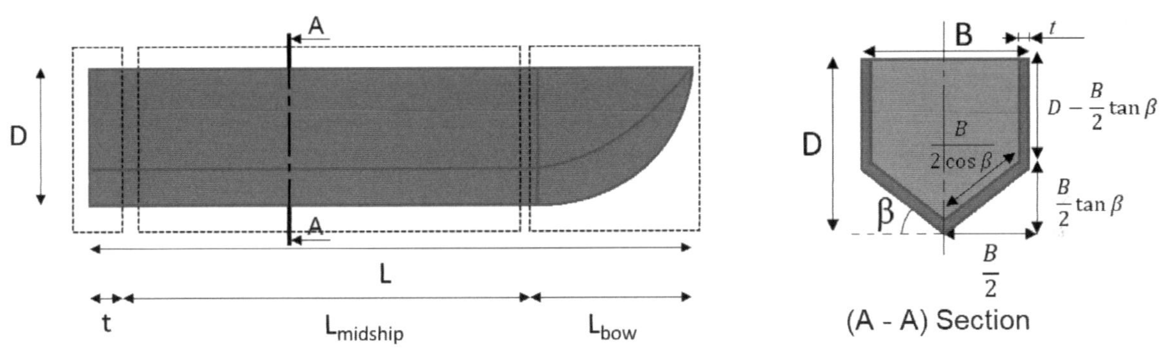

Figure 5: Profile and frontal view of model.

The sum of the transom, mid-ship, and bow sections can then be compounded to produce a total weight estimate for the hull of the high-speed craft.

$$W_{hull} = W_{bow} + W_{mid-ship} + W_{transom} \tag{6}$$

The goal for the weight estimation was to incorporate the process directly into the iterative design code. By developing these estimation methods, modeling the entire hull in SolidWorks for each iteration can be avoided. Once the relationship between bow volume and beam is determined for a given thickness, the MATLAB code can handle the entire hull weight estimation.

Propulsion And Steering

The design of the control and propulsion system faced three key challenges. The propulsion system must be able to propel the craft at speed great enough for the craft to reach planing, the craft must be able to be controlled both by manual input as well as autonomously on a predetermined route, the control and propulsion systems must be modular to allow for further testing with different hull designs.

During the initial stages of the design of the propulsion system the main issue encountered was the model size. The size of the model is considerably larger than most remote-control boats currently on the market, however it is also smaller than any personal watercraft. Three types of motors were initially considered, including trolling motors used on recreational fishing boats, electric motors used to power kayaks and stand-up paddle boards, and large brushless motors used in remote control racing boats. After consideration of ease of application and drag it was determined that motors used in the construction of large-scale remote-control racing boats would most readily fit the application, since both trolling motors and electric motors used in kayaks presented large amounts of drag as well as issues with power and mounting.

For the control system, solutions used on existing remote-control boats available on the consumer market are utilized. The system for manual control is relatively simple with a single rudder that is controlled with a servo as well as a speed controller that will control the speed of the motor. Both the servo and the speed controller will be connected to a 2.4 gHz receiver that will accept inputs from an off the shelf radio controller. The on-board mounted Raspberry Pi will be able to store a predetermined route and through the use of GPS the test craft will be able to conduct test runs without the use of manual inputs. The relative locations of the electronics, servos, and motor are shown in the Figure 6a. Further, a SolidWorks drawing of the rudder is shown in Figure 6b.

The initial design for the propulsion and control system revolves around one initial hull design.

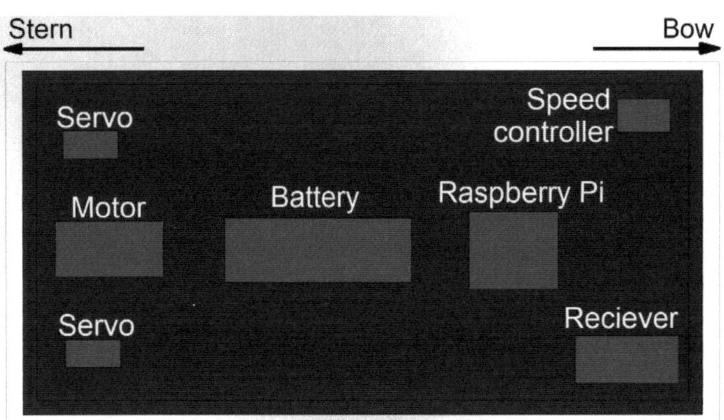

(**a**) Control box.

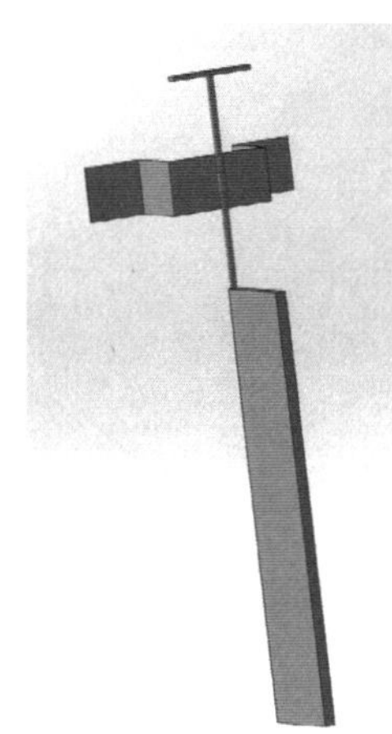

(**b**) Rudder.

Figure 6: Solid works models of propulsion and steering systems

The goal is to be able to use the same components in multiple hulls in order to reduce costs as well as increase efficiency in testing multiple hulls. As a result of this, a self-contained propulsion and control unit was designed revolving around a removable box that will make up a portion of the stern of the hull. This box will be compressed against the hull using a rubber gasket in order to ensure the seal is watertight. This will enable the entire system to be moved from one hull to another without disassembly.

Savitsky's Analysis and Stability Constraints

Savitsky (1964) introduced a simple hydrodynamic empirical model to describe the forces and the equilibrium trim angle at a certain running velocities. His method was applied in a design algorithm to accommodate the free-running test in which the running velocity may vary. This was accomplished by checking the trimming and the longitudinal stability of the iterative hull design at different speeds, as shown in Figure 7 and Algorithm 1 in the Appendix.

The inputs to the trim and porpoising analyses were the model's velocity range, geometry, and weight, proposed during the previous design steps. First, it was checked that the lowest speed in the provided range would ensure planing. Then, the running trim was calculated using the Savitsky (1964) Method, see Appendix: Algorithm 1. This results in either a running trim that cannot be obtained due to the positive moment of the craft at all possible trim angles, or an estimation of the drag force and the running trim. If longitudinal stability could be achieved, the risk of porpoising was calculated using Celano (1998) and Savitsky (1964), see Equations 7a and 7b, respectively. If the calculated equilibrium trim angle was not below the porpoising stability criteria or no trim angle resulted from Savitsky (1964) method, the running velocity was increased, if and only if the trim angle was obtained at the majority (70%) of the velocity range, provided by the motor. If not, a new motor was selected.

$$\tau_{critical,\ Celano} = 0.1197\ \beta^{0.7651}\ e^{15.7132\sqrt{\frac{C_{L_\beta}}{2}}\beta^{-0.2629}} \tag{7a}$$

$$\tau_{Critical,\ Savitsky} = -1.87 + 12.54\ \sqrt{\frac{C_{L_\beta}}{2}} + 80.87\ \frac{C_{L_\beta}}{2} + 0.193\ \beta - 0.0017\ \beta^2 - 0.3125\ \beta\sqrt{\frac{C_{L_\beta}}{2}} \tag{7b}$$

where β is the deadrise angle of hull, and C_{L_β} is the lift coefficient for deadrise surface.

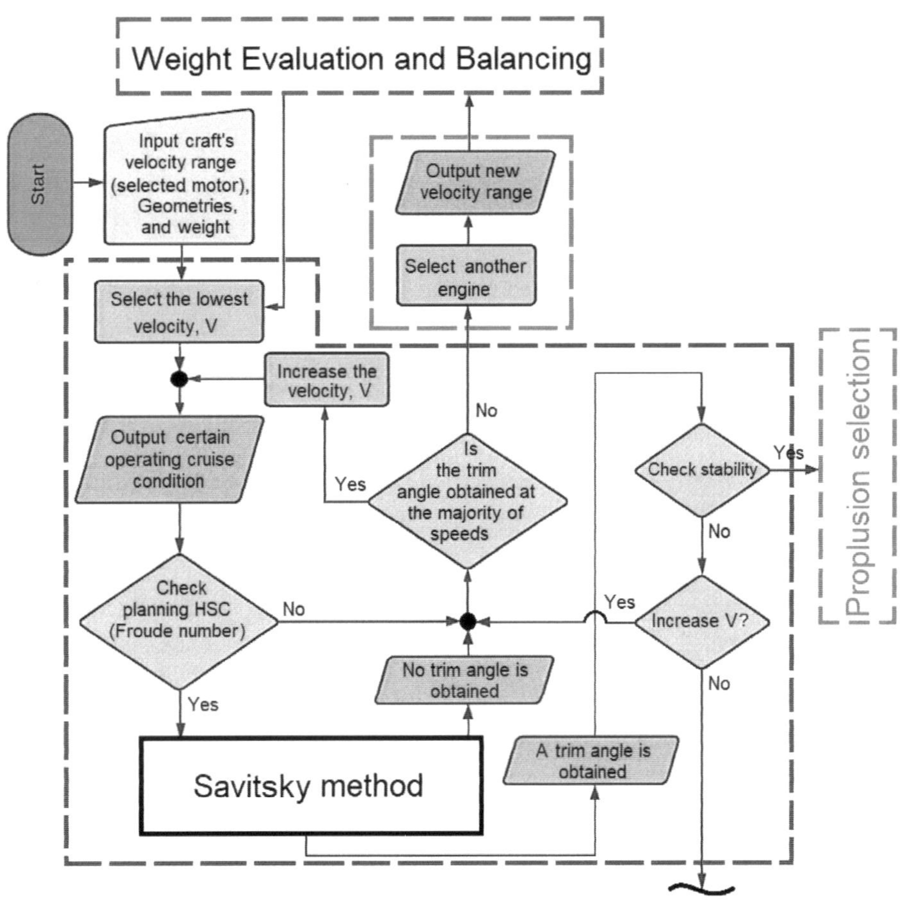

Figure 7: Detailed design workflow for the trim and longitudinal stability of the free-running high-speed model.

Structural Analysis

Using American Bureau of Shipping (July 2020) rules to determine the pressure on the hull, the minimum thickness of the hull can determined, as per American Bureau of Shipping (July 2020) rules. The pressure on the hull is categorized into slamming pressure, p_S, hydrostatic pressure, p_H, and fore-end pressure, p_{FE}. The slamming and hydrostatic pressure exist on the bottom and side parts of the hull. While the fore-end pressure is applied on the bow of the craft. The pressure is calculated using the craft's displacement, Δ, speed, V, waterline length, L_W, waterline breadth, B_W, draught, T, and running trim angle, τ. For simplicity during the early stages of design, the craft's waterline length and breadth are considered the same as the craft's length and breadth ($L_W = L$, and $B_W = B$). Indeed, the bottom design pressure, that is applied to the hull below the chines or the upper turn of the bilge, are the most critical. Thus, our preliminary design will consider the bottom slamming design pressure in calculating the hull's thickness.

Firstly, the estimated bottom slamming design pressure at any section clear of LCG, using the American Bureau of Shipping (July 2020) rules, is expressed in kN/m^2 as:

$$p_{S,bxx} = \frac{0.1\Delta}{L_W B_W}\left[1 + n_{cg}\right] F_D F_V \tag{8}$$

where F_D is the design area factor, which depends on the design area (the actual area of the shell plate panel for plating), A_D, reference area, A_R, and the spacing of longitudinals or stiffeners, s in (mm). Whereas F_V is the vertical acceleration distribution factor which varies along the waterline length of the craft. In the preliminary design, the F_D and F_V are set to their their maximum, $F_D = F_V = 1$, which represents the highest possible values of slamming pressure on the hull's bottom. Later, these factors will be corrected in the detailed design phase with the actual values from the existed tables and figures in the American Bureau of Shipping (July 2020). The vertical acceleration of the craft, n_{cg}, is calculated using the formula of average 1/100 highest vertical accelerations at LCG in terms of the gravitational acceleration, g, during the preliminary stages of design (American Bureau of Shipping July 2020), and expressed as:

$$n_{cg} = 0.0078\left[\frac{12h_{1/3}}{B_W} + 1.0\right]\tau\left[50 - \beta_{cg}\right]\frac{V^2 B_W^2}{\Delta} \tag{9a}$$

where $h_{1/3}$ is the significant wave height and equals to $4m$ for full scale HSC during the operational condition, and β_{cg} is deadrise at LCG (which must be between 10° and 20°). Further, the value of n_{cg} from Equation (9a) should not exceed its maximum limit, which is defined in the American Bureau of Shipping (July 2020) rules as:

$$n_{cg} = 1.39 + \frac{K_V V}{\sqrt{L}} \tag{9b}$$

where K_V is the vertical acceleration distribution factor along the waterline length and its maximum value is 2. In addition, the maximum limit of n_{cg} for speeds greater than $18\sqrt{L}m/s$ is $6.0g$. Further, its typical minimum limit is $2.0g$ for craft whose length is less than $12m$.

Secondly, the bottom hydrostatic design pressure is estimated in the American Bureau of Shipping (July 2020) rules, using Equation 10, in kN/m^2.

$$p_{H,b} = 9.8\left[0.64H + d\right] \tag{10}$$

where H is the wave parameter and equals $(0.0172L + 3.653)m$, but it can not be less than the maximum survival wave height for the craft.

The thickness of the hull must not be less than the critical thickness, given by the American Bureau of Shipping (July 2020) and expressed in Equations 11 and 12 in mm.

$$t_{b,ABS,1} = s\,c^3\sqrt{\frac{p\,k_1}{10\,E_F}} \tag{11}$$

$$t_{b,ABS,2} = s\,c\sqrt{\frac{p\,k}{1000\,\sigma_a}} \tag{12}$$

where p is the design pressure, c is the plate curvature factor and its value is one for our HSC model, σ_a is the design stress and it equals to 0.33 of the minimum flexural strength, σ_u in N/mm^2, E_F is the flexural modulus in N/mm^2, and k and k_1 are coefficients which vary with the plate panel aspect ratio. The maximum values of these coefficients are 0.5 and 0.028 at the highest plate panel aspect ratio, respectively.

Selection of Main Particulars

The design procedure, shown in Figure 1, was tested for different length to breadth ratios between the given limits by the statistical analysis, ($2.5 < L/B < 4.5$). The length of the craft was fixed to $2.16m$ ($85in$), see Design constraints section. In addition, the remaining ship's parameters, B, T, and V were calculated based on the statistical data of all HSC (Black line in Figures 2, and 3). Further, the Foam FR-4615 material was chosen, and its density, Flexural Strength, and Flexural Modulus are $240kg/m^3$, $4.94KPa$, and $156.85KPa$, respectively. The thickness of the hull started at $2.54cm$ ($1in$) and changed to accommodate the applied pressure while staying above the critical thickness, calculated using the American Bureau of Shipping (July 2020) rules. The performed structural analysis showed that the thickness needed to be increased to $3.5cm$ ($1.378in$) to withstand the applied pressure. In addition, the weight estimation of both the electronics and the motor was estimated at $1.15kg$.

The selected ratios to be evaluated were $L/B = 3$, 3.25, 3.5, 3.75, and 3.95. Figure 8 presents the variation of the craft's beam, mass, and the maximum achievable velocity (that provides a trim angle), estimated by the Savitsky (1964) Method. It was noticed that the craft's mass decreases, and the maximum applicable velocity increases with the increase in length to breadth ratio, due to the decrease in beam. Furthermore, the calculated equilibrium trim, shown in Figure 9, must be below the critical trim where porpoising occurs, defined in section Savitsky's analysis and stability Constraints. The gray color in the figure's legend is used for the line types to specify different trim angles; critical porpoising trim angles are shown in blue (Celano 1998) and red (Savitsky 1964), and equilibrium trim angle is in black. The type of the gray-colored line specifies the length to breadth ratio. Further, the green line represents the starting limit of planing crafts, where SLR is larger than 3 (Savitsky 2003). The yellow line shows the maximum velocity to obtain a trim angle for a stable HSC model. A higher velocity would result in positive moment with a risk for overturning of the model, i.e. no equilibrium trim angle exists. Hence, the designed HSC model is (theoretically) porpoising stable, and not necessarily is the real case because the difference between the running trim angle and the critical ones is within a degree. Thus, a further computational study is needed before the manufacture phase and followed by a free-running test in small body of water, such as a lake.

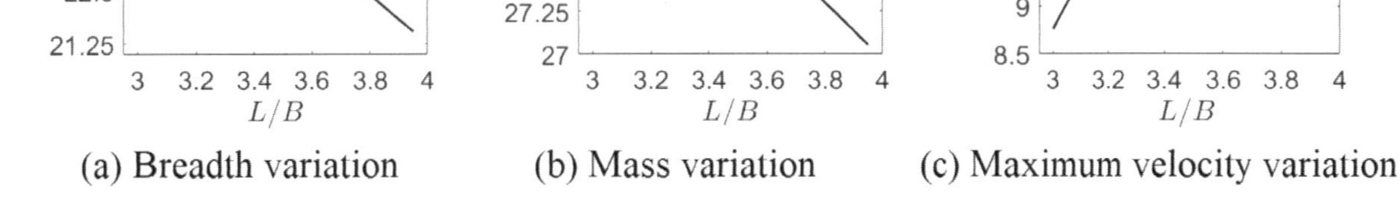

(a) Breadth variation (b) Mass variation (c) Maximum velocity variation

Figure 8: Breadth, mass and maximum velocity as a function length to Breadth ratio .

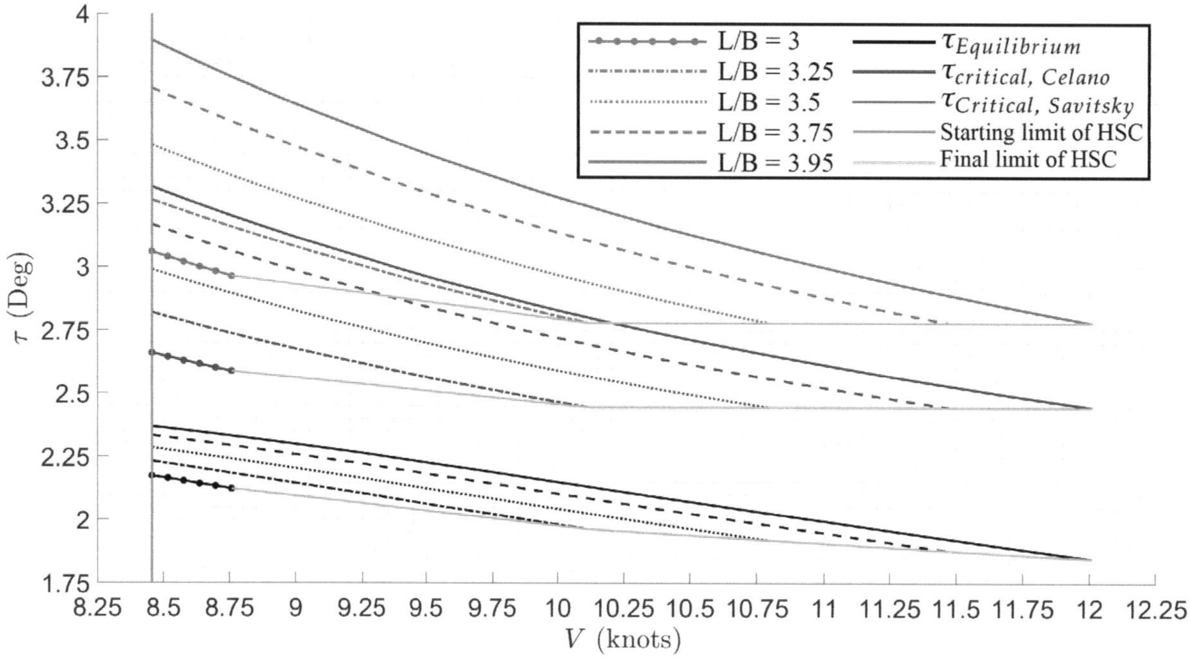

Figure 9: Calculated and critical porpoising trim (Celano (1998) and Savitsky (1964)) for different *L/B* ratios. Critical porpoising trim angles are shown in blue (Celano 1998) and red (Savitsky 1964), and equilibrium trim angle is in black. The type of the gray-colored line specifies the length to breadth ratio.

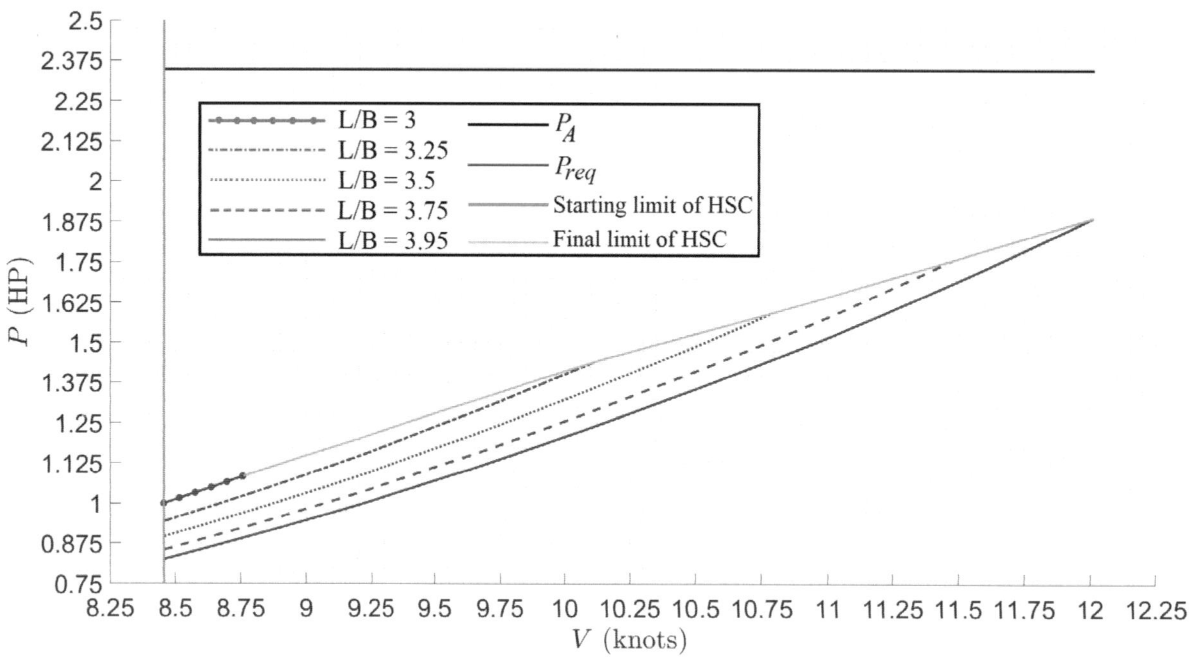

Figure 10: Required power as a function of speed for different L/B ratios compared to the available power by the selected motor (black solid line). Critical porpoising trim angles are shown in blue (Celano 1998) and red (Savitsky 1964), and equilibrium trim angle is in black. The type of the gray-colored line specifies the length to breadth ratio.

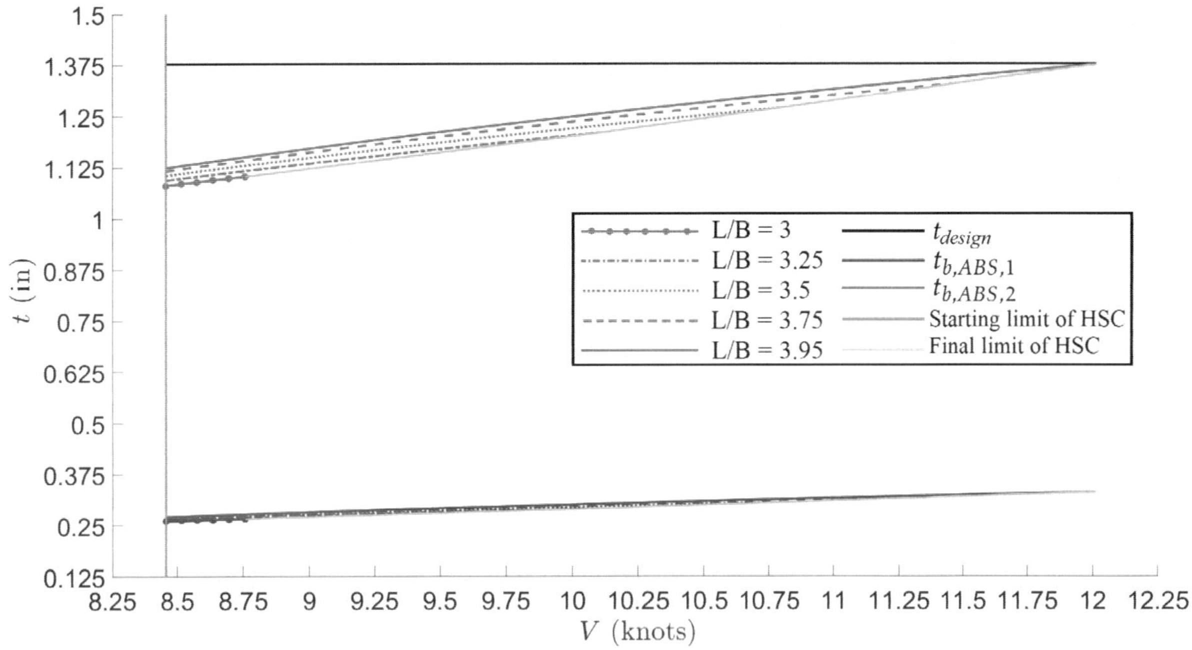

Figure 11: Needed minimum thickness (American Bureau of Shipping (July 2020)) as a function of speed for different L/B ratios. Critical porpoising trim angles are shown in blue (Celano 1998) and red (Savitsky 1964), and equilibrium trim angle is in black. The type of the gray-colored line specifies the length to breadth ratio.

In addition, a comparison was conducted between the required power, P_{req} to overcome the drag force and drive the HSC model, and the available power supplied by the motor, as shown in Figure 10, to ensure the feasibility of the selected motor. The performed structural analysis, shown in Figure 11, showed that the final selected thickness $3.5cm$ ($1.378in$) will withstand the applied pressures on the hull for length to breadth ratios less than 3.95. Hence, a length to breadth ratio of 3.75 was selected.

Computer Aided Design

The model is a V-shaped prismatic hull. The craft was modelled in SolidWorks, using the obtained design values during the design procedure. A model of the craft is shown in Figure 12 whereas the parameters for the craft are given in Table 1.

Parameter	Symbol	Imperial Unit	SI Unit
Length overall	Loa	$85in$	$2.16m$
Waterline overall length	Lwl	$75.2in$	$1.91m$
Breadth	B	$22.677in$	$0.576m$
Draught	T	$15in$	$0.38m$
Deadrise angle	β	$20°$	$20°$
Thickness	t	$1.378in$	$0.035m$
Mass	M	$27.381lbs$	$12.42kg$
Longitudinal center of gravity from transom along keel line	LCG	$25.591in$	$0.65m$
Vertical center of gravity above keel line	VCG	$7.874in$	$0.2m$

Table 1: Model characteristics

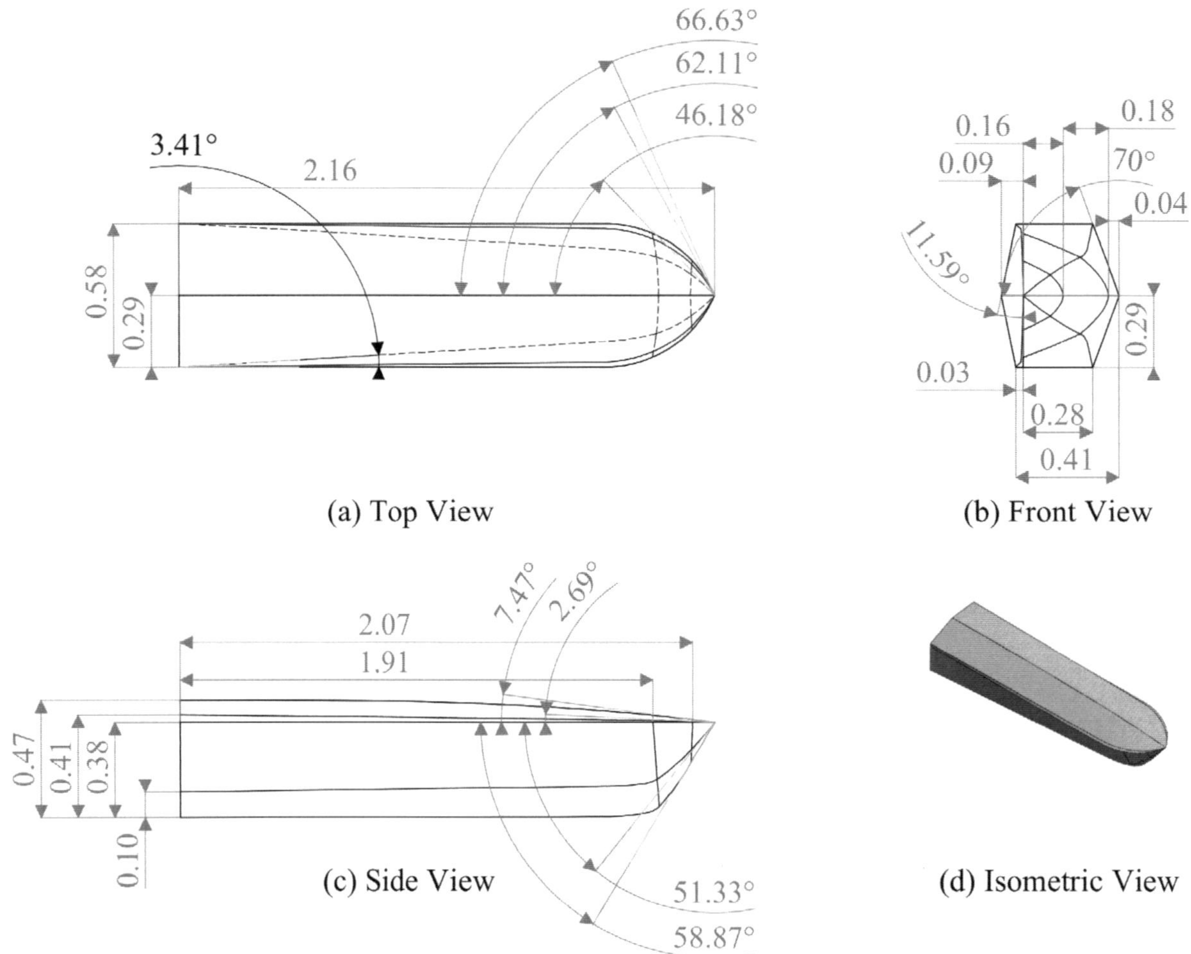

Figure 12: High Speed Craft SolidWorks model

Testing Site Selection

In the convectional HSC towing tank testing, a significant wave height to hull beam ratio, $h_{1/3}/B$, ranges between 0.22 to 0.66 (Fridsma 1971), however, higher end of this range could be severe for free running tests where the hull is unconstrained (Wang et al. 2020). Therefore, based on the narrower and less severe range chosen by Wang et al. (2020), a range of $[0.0864 - 0.144]m$ was selected based on a $B = 0.576m$. Recorded wave height data by wave buoys of the National Oceanic and Atmospheric Administration (NOAA) at the Gulf of Mexico, Earle (1996), shows that the significant wave height operating range can be achieved. Four wave buoys will be installed at the testing location that will provide a real-time monitoring for the sea-state conditions to ensure that the surrounding environment within the allowable range and to relate our results to the measured the significant wave heights and the significant wave periods in the testing location.

Although the experiment will be conducted off the coast at the Texas A&M Galveston campus, the design, manufacturing and testing will take place at the Texas A&M College Station Campus. An initial testing will be performed in Bryan City Utilities Lake that is 11.3 miles from the College Station campus. Lake Bryan has very calm water conditions and quite moderate depth as shown in the bathymetric contour Figure 13, and this will help to initially test the HSC movements, control, data monitoring and test the installation of the wave buoys and its mooring system.

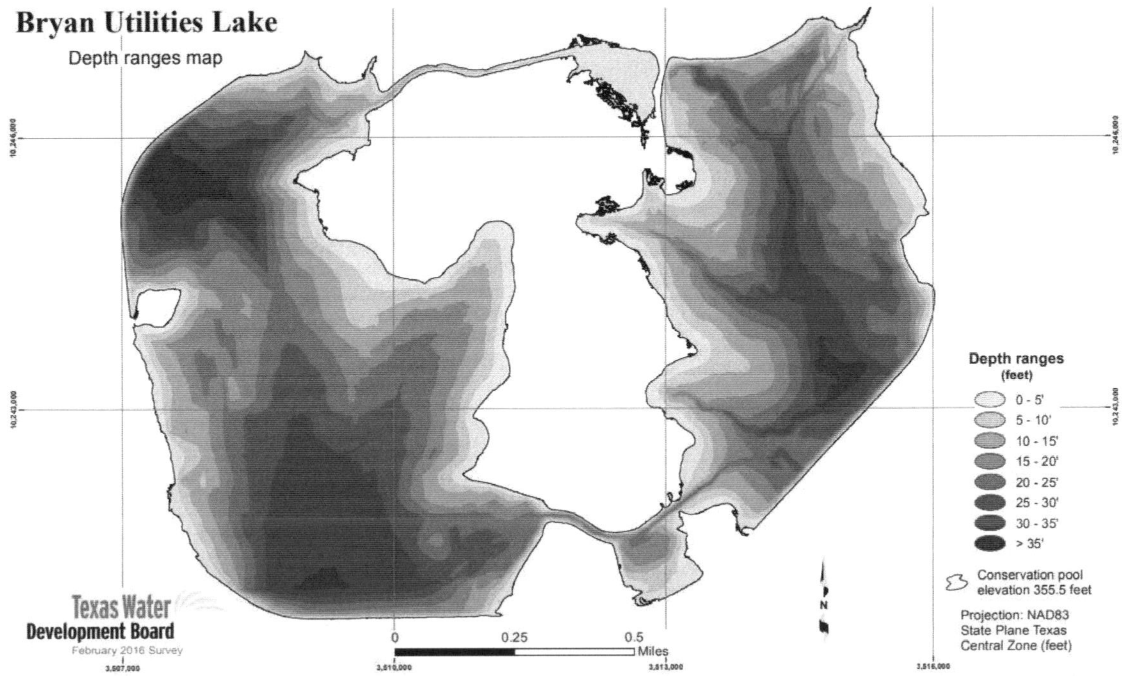

Figure 13: Bathymetric map for Bryan utilities lake , Kemp et al. (2016)

System Monitoring And Data Acquisition

In order to assess the total dynamic performance of the hull during the free running test; roll, pitch and yaw motions' time series in addition to their corresponding angular velocities and accelerations will be measured. In addition, measuring the applied forces acting on the hull body, such as the slamming and drag forces will enable a comparative analysis between different hull designs to facilitate better designs. Selection of the the HSC motion dynamics measurement system is discussed in this paper, however the pressure slamming forces and drag forces measurements will be presented as a part of the future work.

Inertial Navigation System (INS)

To monitor the roll, pitch and yaw motions, a dual compass Inertial Navigation System (INS) by Inertial-Sense is chosen Sense (2021). This INS development kit is a 9 Degree of Freedom (DOF) Inertial Measurement Unit (IMU) that computes the 3 motions precisely based on the sensor fusion algorithms that combine readings of the three embedded sensors (accelerometer , magnetometer and gyroscope). This INS development kit is a dual compassing system that, beside the magnetometer, is supported by a dual Global Navigation Satellite System (GNSS) receiver that allows the estimation of the hull heading (yaw) without reliance on the magnetometer readings (that may be affected by stationary magnetic noise generated by the surrounded ferrous metals or the dynamic magnetic noise by the rotating motors magnetic field). This INS development kit has an accuracy 0.1° for pitch and roll motions and 0.3° for the heading, and data output rate up to 1 KHz and 0.5 KHz for the raw and processed data respectively. As the HSC Length overall L_{oa} is only 2.16 m and the hull will be manufactured from brittle materials, for this initial design the body will be assumed to be rigid. Hence, a single IMU sensor will installed on-board to monitor the hull movements and accelerations.

Real-Time Kinematic-Global Navigation Satellite System (RTK-GNSS)

Measuring the model speed over ground or controlling the hull trajectory on a predetermined route requires precise real time location monitoring. By the use of Real Time Kinematic (RTK) technology Dominguez et al. (2010) available in the Inertial-Sense development kit we will be able to enhance the on board dual GNSS position estimation to reach 1-3 cm accuracy compared to the traditional GNSS that has an accuracy 2-5 m. The RTK-GNSS consists of 2 GNSS receivers (stationary and rover) the rover is installed on the hull and the stationary one is installed onshore, both receive signals from the satellites and communicate with each other for position error correction, as illustrated in Figure 14. The stationary and the rover GNSS receivers has different four frequency bands that enable the system to receive signals from four different GNSS (GPS, GLONAS, Galileo and BeiDou) that helps to maintain and stabilize a continuous position reading.

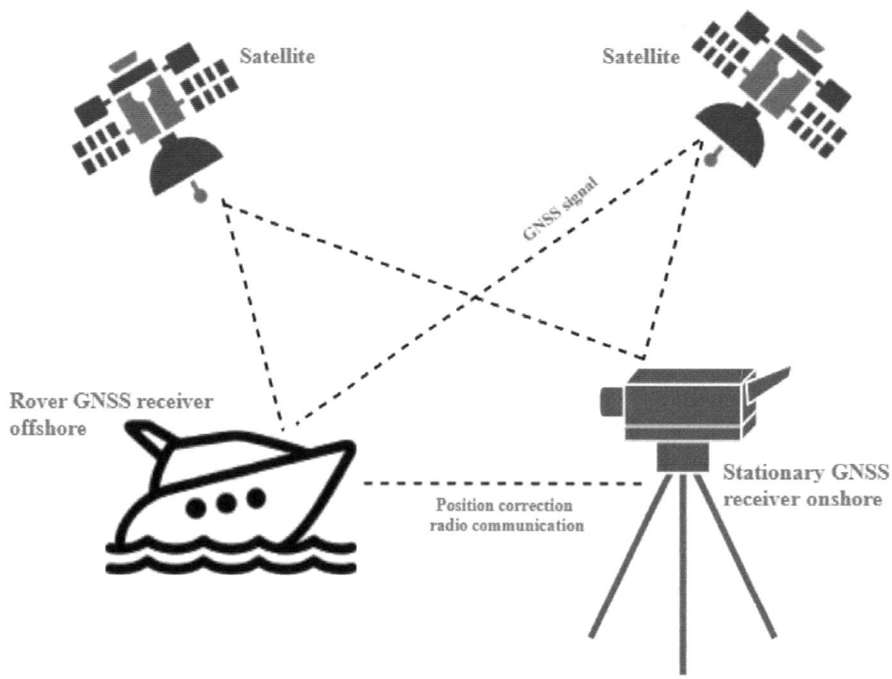

Figure 14: Schematic diagram of the RTK-GNSS for the HSC position monitoring

Summary

In this paper a step-by-step workflow is presented for the design of a High-Speed Craft (HSC) for open sea free running tests in the Gulf of Mexico off the coast of Texas A&M Galveston Campus. First, a comprehensive statistical data analysis were performed for a wide range of existing successful HSC with different usage areas, these statistics include main non-dimensional design parameters such as the variation of Froude number with length to beam ratio, the variation of beam to draught ratio with length to beam ratio and the variation of weight to length-beam-draught ratio with length to beam ratio. Next, a design spiral algorithm was developed to determine a suitable overall length, beam, depth, and mass displacement. Further, a suitable running trim and Longitudinal Center of Gravity that does not risk porpoising was determined. The paper also discussed possible material selections for the hull manufacturing, along with the propulsion and steering systems' design. Finally, sensors for system motioning are selected to accurately track the hull free motion dynamics, position, and the applied hydrodynamic forces.

Future Work

The design of an HSC open-sea free running model is still in the initial stages. The next step is to prepare the CAD for manufacturing purposes. Although the hull resistance and the propulsion power are estimated in the design workflow, these estimations will be verified by a two-phase Computational Fluid Dynamics (CFD) simulation using OpenFOAM. Once the hull manufacturing is done, the system monitoring and data acquisition system will be installed and initially tested in the Bryan Utilities Lake. Other future work plans are summarized as below:

- Update the used ABS structural analysis to include the longitudinals and stiffeners.
- Create a detailed CAD model of the designed HSC with longitudinals, stiffeners, deck, bulkhead plating, covering shell, and the precise equipment's weights and locations.
- Perform a numerical hydrodynamic simulation for the designed model using OpenFOAM, to investigate the vehicle behavior in a free-running test and calculate the applied hydrodynamic forces on the hull.
- Perform a numerical stress analysis for the hull structure, using the hydrodynamic forces obtained in the OpenFOAM simulation.
- Determine the manufacturing process and material of the hull
- Based on the design planning speed and the predicted hydrodynamic drag forces, a comprehensive study will be conducted to design and select the propulsion and powering systems including (batteries' capacity, brushless motor power and rpm characteristics, propellers' pitch and diameters).
- Perform an experimental correlation between the propulsion system thrust force and the power consumption to be able to monitor the applied drag forces on the HSC during the free-running test.
- Conduct forward and turning movements experiments in the Bryan lake to evaluate the vehicle movements transfer functions, then design controllers for the autonomous path tracking algorithm.

Acknowledgements

This work was sponsored by the Office of Naval Research, ONR award number N00014-20-1-2855, Robert Brizzolara is the program manager.

Nomenclature

A_D	Design Area
A_R	Reference Area
B	Craft's Breadth
B_W	Waterline Breadth
c	Plate Curvature Factor
C_b	Block Coefficient
C_{L_β}	Lift Coefficient for Hull Deadrise
E_F	Flexural Modulus
F_D	Design Area Factor
F_V	Vertical Acceleration Distribution Factor
$h_{l/3}$	Significant Wave Height
K_V	Vertical Acceleration Factor Along the Waterline
L	Craft's Length
L_W	Waterline Length
M_{HSC}	High speed craft's Mass
n_{cg}	Vertical Acceleration of the craft
p	Design Pressure
P_A	Available Power of Motor
p_H	Hydrostatic Pressure
p_S	Slamming Pressure
p_{FE}	Fore-end Pressure
s	Spacing of Longitudinal Stiffeners
T	Craft's Draught
t	Hull's Thickness
W	Craft's Weight
W	Required Power for Craft
β	Hull Deadrise
β_{cg}	Deadrise at LCG in Degrees
Δ	Craft's Displacement

∇	Displaced Volume of Water
ρ	Density
σ_a	Design Stress
σ_u	minimum flexural strength
$\tau_{critical}$	Critical Trim
ABS	American Bureau of Shipping
CFD	Computational Fluid Dynamics
DOF	Degree of Freedom
FSI	Fluid Structure Interaction
$GNSS$	Global Navigation Satellite System
HSC	High Speed Craft
IMU	Inertial Measurement Unit
INS	Inertial Navigation System
KCS	KRISO Container Ship
LCG	Longitudinal Center of Gravity
$LORCA$	Lehigh Ocean Research Craft Autonomous
RTK	Real Time Kinematic technology
SLR	Speed Length Ratio
USV	Unmanned Surface Vehicle
VCG	Vertical Center of Gravity

References

American Bureau of Shipping. Rules for building and classing high speed craft. part 3: Hull construction and equipment, July 2020.

Luther H Blount. Hull construction, December 28 1982. US Patent 4,365,580.

Morten Breivik, Vegard E Hovstein, and Thor I Fossen. Straight-line target tracking for unmanned surface vehicles. *Modeling, Identification and Control Journal*, 29(4):131–149, 2008.

Tullio Celano. The prediction of porpoising inception for modern planing craft. *SNAME Trans*, 106:269–292, 1998.

Eugene P Clement. Effects of longitudinal bottom spray strips on planing boat resistance. Technical report, DAVID TAYLOR MODEL BASIN WASHINGTON DC, 1964.

Brock Cooper. Differences between carbon fiber & fiberglass, Mar 2019. URL https://sciencing.com/differences-between-carbon-fiber-fiberglass-8267799.html.

Sujit Das. Life cycle assessment of carbon fiber-reinforced polymer composites. *The International Journal of Life Cycle Assessment*, 16(3):268–282, 2011.

Benedicto H Dominguez, Thomas J Manessis, Janet T Roth, Pascal Achille Caillon, Jason Spielman, Terence Spielman, and James Donald Reno. Mobile account authentication service, April 27 2010. US Patent 7,707,120.

Marshall D Earle. Nondirectional and directional wave data analysis procedures. *NDBC Tech. Doc*, 96(002), 1996.

Soundings Editors. Boat building for the rest of us. part ii, Oct 2007. URL https://www.soundingsonline.com/news/boatbuilding-for-the-rest-of-us-part-ii.

Gerard Fridsma. A systematic study of the rough-water performance of planing boats. irregular waves-part 2. Technical report, STEVENS INST OF TECH HOBOKEN NJ DAVIDSON LAB, 1971.

Stefano Gaggero, Diego Villa, and Michele Viviani. An extensive analysis of numerical ship self-propulsion prediction via a coupled bem/rans approach. *Applied Ocean Research*, 66:55–78, 2017.

Eric Greene. Marine composites. annapolis: Eric greene associates, 1999.

Joachim Grenestedt, Jack Keller, Scott Larson, Jake Patterson, John Spletzer, and Tyler Trephan. Lorca: A high performance usv with applications to surveillance and monitoring. In *2015 IEEE International Symposium on Safety, Security, and Rescue Robotics (SSRR)*, pages 1–6. IEEE, 2015.

Quality Systems Group 28th ITTC, editor. *Free Running Model Tests*, volume Procedure 7.5-02-01-01, 2014. International Towing Tank Conference (ITTC).

Jialong Jiao, Shuzheng Sun, Jide Li, Christiaan Adika Adenya, Huilong Ren, Chaohe Chen, and Dongjiao Wang. A comprehensive study on the seakeeping performance of high speed hybrid ships by 2.5 d theoretical calculation and different scaled model experiments. *Ocean Engineering*, 160:197–223, 2018.

Jason J. Kemp, Holly Holmquist, Khan Iqbal, and Nathan Leber. Volumetric survey of bryan utilitie lake. Technical report, Texas Water Development Board, 2016.

Shafi Ullah Khan, Chi Yin Li, Naveed A Siddiqui, and Jang-Kyo Kim. Vibration damping characteristics of carbon fiber-reinforced composites containing multi-walled carbon nanotubes. *Composites science and technology*, 71(12):1486–1494, 2011.

M Lakatos, T Sahk, R Kaarma, K Tabri, M Kõrgesaar, and H Andreasson. Experimental testing of spray rails for the resistance reduction of planing crafts. In *Trends in the Analysis and Design of Marine Structures: Proceedings of the 7th International Conference on Marine Structures (MARSTRUCT 2019, Dubrovnik, Croatia, 6-8 May 2019)*, page 334. CRC Press, 2019.

Edward M Lewandowski. Free running model turning tests of the us coast guard 47 ft motor life boat. Technical report, STEVENS INST OF TECH HOBOKEN NJ DAVIDSON LAB, 1995.

Bogdan Molchanov, Svante Lundmark, Mirjam Fürth, and Mathew Green. Experimental validation of spray deflectors for high speed craft. *Ocean Engineering*, 191:106482, 2019.

Zuzana Murčinková, Imrich Vojtko, Michal Halapi, and Mária Šebestová. Damping properties of fibre composite and conventional materials measured by free damped vibration response. *Advances in Mechanical Engineering*, 11(5):1687814019847009, 2019.

Linus Olin, Mireia Altimira, Jonas Danielsson, and Anders Rosén. Numerical modelling of spray sheet deflection on planing hulls. *Proceedings of the Institution of Mechanical Engineers, Part M: Journal of Engineering for the Maritime Environment*, 231(4):811–817, 2017.

Daniel Savitsky. Hydrodynamic design of planing hulls. *Marine Technology and SNAME News*, 1(04):71–95, 1964.

Daniel Savitsky. On the subject of high-speed monohulls. *Greek Section of the Society of Naval Architects and Marine Engineers (SNAME)*, 2003.

Inertial Sense. Dual compassing l1/l2 multi-band gps with ins development kit + rtk capable, Jan 2021. URL url={https://inertialsense.com/product/dual-gps-l1-l2-multi-band-ins-rtk-development-kit/}.

Jeonghwa Seo, Hak-Kyu Choi, Uh-Cheul Jeong, Dong Kun Lee, Shin Hyung Rhee, Chul-Min Jung, and Jaehoon Yoo. Model tests on resistance and seakeeping performance of wave-piercing high-speed vessel with spray rails. *International Journal of Naval Architecture and Ocean Engineering*, 8(5):442–455, 2016.

Zhirong Shen, Pablo M Carrica, and Decheng Wan. Ship motions of kcs in head waves with rotating propeller using overset grid method. In *International Conference on Offshore Mechanics and Arctic Engineering*, volume 45400, page V002T08A043. American Society of Mechanical Engineers, 2014.

Abhinav Valada, Prasanna Velagapudi, Balajee Kannan, Christopher Tomaszewski, George Kantor, and Paul Scerri. Development of a low cost multi-robot autonomous marine surface platform. In *Field and service robotics*, pages 643–658. Springer, 2014.

KD von Ellenrieder. Free running tests of a waterjet propelled unmanned surface vehicle. *Journal of Marine Engineering & Technology*, 12(1):3–11, 2013.

Xinguo Wang, Jack Bonoli, Madeline Cohen, and Mirjam Fürth. Design and testing outline for a free running model of a high speed craft. In *HSMV 2020: Proceedings of the 12th Symposium on High Speed Marine Vehicles*, volume 5, page 21. IOS Press, 2020.

Chiara Wielgosz, M Fürth, R Datla, U Chung, Anders Rosén, and J Danielsson. Experimental validation of numerical drag prediction of novel spray deflector design. In *13th International Marine Design Conference, IMDC 2018, 10 June 2018 through 14 June 2018*, pages 491–497. CRC Press/Balkema, 2018.

Dimitrios-Alexandros Zisimopoulos. *Use of Fiber Reinforced Plastics in Ship Construction: A Study of SOLAS regulation II-2/17 on Alternative Design and Arrangementsfor Fire Safety.* PhD thesis, National Technical University of Athens, 2016.

Appendix

Algorithm 1: Pseudo-Code to calculate the trim angle at certain running velocity using Savitsky (1964) approach.

Input : Hull's geometry (β, L, and B), velocity (V), estimated weight (Δ), and center of gravity location (LCG, and VCG).

Output: Check the applicability of the selected velocity for the designed HSC model and calculate the equilibrium trim angle (τ_{equil}).

1 Calculate the lift coefficient for deadrise surface (C_{L_β}).
2 Calculate the lift coefficients for zero deadrise surface (C_{L_o}).
3 Start with a very small assumed trimming angle ($\tau \geq 0$).
4 Calculate the speed coefficient (C_v).
5 Calculate the mean wetted length-beam ratio (λ).
6 Calculate the wetted keel length (L_k).
7 **while** *Wetted keel length is larger than the hull's length* ($L_k > L$) **do**
8   Increase the trimming angle with small value ($\tau = \tau + \delta\tau$)
9   Recalculate steps 5, and 6.
10 **end**
11 Calculate the mean velocity over bottom of planing surface (V_1).
12 Calculate the Schoenherr turbulent-friction coefficient (C_f).
13 Calculate the hydrodynamic drag due to the turbulent-friction (D_f).
14 Calculate the distance of center of pressure (C_p).
15 Calculate the distance between the normal component of resistance force and center of gravity (c).
16 Calculate the pitching moment (M_x).
17 **if** *Pitching moment is a positive value* ($M_x > 0$) **then**
18   The running velocity (V) IS NOT applicable; Change it.
19 **else**
20   The running velocity (V) IS applicable.
21   **while** *Pitching moment is a negative value* **do**
22   Increase the trimming angle with small value ($\tau = \tau + \delta\tau$)
23   Recalculate steps 5, and 11 - 16.
24   **end**
25 **end**
26 Interpolate between the last two last iterations to get the equilibrium trim angle (τ_{equil}), and the other parameters at the equilibrium condition.

AN OPEN-SOURCE FRAMEWORK FOR SHIP PERFORMANCE CFD

Björn Windén
Founder, SHORTCUT CFD

Proceedings of the 26th Offshore Symposium, April 6th & 7th 2021, Virtual, Houston, Texas
Texas Section of the Society of Naval Architects and Marine Engineers

ABSTRACT

Development of novel hydrodynamic solutions to meet the demands for fuel efficiency of the future merchant fleet will require continuous innovation in the Computational Fluid Dynamics (CFD) methods being used. This is especially true when considering the development of Energy Saving Devices (ESDs). Open-source CFD is a way to allow more researchers to participate in this process and to accelerate innovation. This paper presents an open-source framework for conducting ship performance analyses within the OpenFOAM CFD suite. The framework helps to alleviate the steep learning curve for initially setting up a successful model of a self-propelled ship in OpenFOAM. It is demonstrated that, using this framework, results in line with the current state of the art set by leading institutions and commercial solvers can be obtained for the performance of a bulk carrier fitted with an ESD. Two different models for describing the action of the propeller on the fluid are explored and conclusions are drawn from comparisons with available experimental data.

Keywords: CFD, Open-Source, Ship Performance, Propeller, ESD, OpenFOAM, SHORTCUt

Introduction

The maritime industry is facing ever-increasing demands for the efficiency of the world's merchant fleet. This is illustrated by the International Maritime Organization (IMO) introducing the Energy Efficiency Design Index (EEDI) as a mandatory design constraint for new ships from 2013 (IMO, 2011). EEDI-related constraints are staggered over time and by 2025, they will come to mandate a 30% reduction in CO_2 emissions from shipping compared to a 2010 baseline. By 2050, the demand will be for a 50% reduction. If the number of vessels in the merchant fleet is assumed to grow at the current rate, a total reduction by up to 85% for individual ships may be required to meet this target (Hesketh & Wittels, 2020).

Traditional hull- and propeller shapes had already reached a relatively advanced stage of optimization for the ships in the baseline generation. This means that traditional ships, powered using traditional fossil fuels, will soon become unable to meet international regulations. Therefore, designers and operators have been forced to come up with innovative ways to reduce the emissions of new ships. Examples include slow steaming, novel fuels and Energy Saving Devices (ESDs).

Slow steaming significantly reduces the emissions of faster ships due to the steep increase in wave-making resistance for high Froude Numbers. However, it has a lower impact for ships such as bulk carriers and tankers which already operate at slower speeds. Slow steaming may also be difficult to justify economically when it is not combined with a global economic crisis that reduces transportation demand and increases the supply of idle vessels (Carson, 2012).

Novel fueling options such as LNG, hydrogen, biodiesels, electric and hybrid options show promise when it comes to meeting the demands of the future merchant fleet (Hesketh & Wittels, 2020). However, many of these options currently lack practical applicability to large ocean-going vessels in terms of safety, reliability and size constraints. Therefore, more conventional fossil fuels will likely have to be in continued use for some time. This is time that ship operators wanting to comply with regulations do not have.

Therefore, much effort has been focused towards finding increased propulsion efficiency through ESDs (such as ducts, fins, novel hubs or rudders) and lower overall thrust requirement through Energy Saving Methods such as air lubrication, kites and sails. An effective ESD would lower the powering requirement for a certain ship. Current methods range from around 3-7% reduction in the powering requirement (ITTC, 2017a). Not only would this reduce the CO_2 emitted by a ship powered by conventional fossil fuels. It would also increase the viability of using some novel fuels which currently have a much lower power density than conventional fuels (and would take up too much space if the powering requirement is too high.) There are many types of ESDs, but for the purpose of this paper, a type used to increase propeller efficiency will be addressed. The successful design of such ESDs relies heavily on detailed hydrodynamic analysis of the ship wake and propeller/hull interaction.

Over a century of experience with ship performance analysis using model tests have resulted in trusted methods to correlate model scale results with full (actual) scale powering requirement. The 1978 powering prediction method recommended by the International Towing Tank Conference (ITTC) is widely used (ITTC, 2017b). However, the differences in wake shape between model scale and full scale mean that appendages designed by looking at model scale flow may not be as efficient when operating in the full-scale wake (Visonneau et.al., 2016). Subsequently, it has been recognized that currently available correlations do not work well for ships with unconventional appendages, such as ESDs. Attempts to correct for this have been made, (ITTC, 1999; 2017a) but have not been able to be representative of a diverse range of devices (Kim et.al., 2017).

One way to overcome these problems is to analyze the viscous flow around the hull using Computational Fluid Dynamics (CFD). This can be done in both model- and full-scale and can therefore be used to study the scale effects of ESDs. CFD also promises to deliver an entirely new form of performance analysis where the hydrodynamic analysis is conducted directly in full scale without the need for a physical model. However, the fidelity of such analyses is limited by the lack of full-scale validation data from ships during normal operation. More detailed data has recently become available (Hiroi et.al., 2019; Sakamoto et.al., 2020) but studies are still few and far between. Nevertheless, advances are being made in CFD development and its application to the analysis of ESD performance. This promises to help bridge the gap between conventional and novel fuels and allow the world to be on track to meet the necessary reductions in CO_2 emissions.

WHY OPEN-SOURCE?

A major advantage of CFD over model testing is its accessibility for a wider range of researchers. Traditional model tests require costly facilities and a large amount of capital investment. In terms of hardware, CFD theoretically only requires computing power; a resource that has grown exponentially for decades and is available to anyone with an Internet connection or even a decent workstation. Having more researchers participate in the hydrodynamic analysis of ship wakes and ESDs will help to bring in new ideas, and to refine existing ones by increasing the amount of available data.

However, one obstacle for smaller entities and individuals wanting to use CFD is still the cost, especially in terms of scalability. Meeting modern CFD standards often necessitates the use of High-Performance Computing (HPC) which can quickly drive up the cost of using software that charges more for each processor core used.

Free and open-source CFD suites such as OpenFOAM not only make more financial sense for smaller businesses and institutions. They also allow a community-driven approach to CFD development which can accelerate innovation while also bringing a unique set of challenges (Herbsleb et. al., 2015; Marsan et.al., 2018). Accelerated innovation is something that has greatly benefitted the industry by allowing users, researchers and developers, to share ideas and new methods. The synergy effect of open-source CFD will be greater when more people participate.

Ship performance analysis requires a high level of complexity when it comes to the hydrodynamics involved, but also when it comes to the geometry itself. This could deter inexperienced maritime industry users from adding CFD analysis to their toolset. Therefore, any solutions to overcoming the sometimes-steep learning curves of open-source marine CFD will help to make CFD hydrodynamics analysis available to a wider range of potential users in the maritime industry; and thereby benefit the industry as a whole.

OVERVIEW

In this paper, an open-source framework for conducting ship performance analysis in OpenFOAM (OpenFOAM Foundation, 2019) is presented. The package, named SHORTCUt, is freely available for download (SHORTCUt CFD, 2021).

The framework has two main parts:

- A pre-processor for OpenFOAM that allows the user to easily set up a ship performance analysis. This replaces the dozens of settings files used in standard OpenFOAM with a single one; featuring terms familiar to Naval Architects, such as Froude Number (Fn), Length between perpendiculars (Lpp) etc.
- An addon library for OpenFOAM, named selfPropFoam, that allows for fast, modular, flexible and accurate propeller modelling without the need for complex meshing.

The selfPropFoam libraries come with the SHORTCUt package but can also be downloaded as standalone for users who do not need the pre-processor.

PRE-PROCESSOR FOR OPENFOAM SHIP PERFORMANCE ANALYSIS

OpenFOAM operates by reading settings from text files. When one wants to pre-process, mesh, run and post-process a case involving complexities such as propeller modeling, a large number of settings files in different folders will be used. For example, the boundary condition for each variable is saved in a separate file. If the user wants to modify the inflow speed, geometry, mesh quality or any other conditions, care must be taken that all files are updated accordingly.

The SHORTCUt pre-processor is designed to streamline this process. Settings that are suitable for a generic ship performance analysis are copied from a built-in template. This is modified using certain key settings that describe a particular ship, propeller and running condition. These key settings are saved in a single file called settingsDict. When using SHORTCUt, this is the only file that needs to be modified to obtain a certain calculation setup. This makes it easy to share an entire calculation setup by simply sharing a single text file. Some key features of the pre-processor and settingsDict are listed below:

- Parameters familiar to Naval Architects (Fn, Lpp etc.)
- Parametric meshing with a single scale factor to control density. Useful for mesh independence studies.
- Automatic adjustments of near-wall mesh for varying Reynolds Numbers
- Support for up to 10 appendages in separate CAD-files.
- Automatic calculation of forces and propulsion coefficients after finished simulations.

Apart from easy case-sharing, another benefit of using a templated setup is that comparative studies are easier to make. Below are some examples of studies that can be conducted by varying a single parameter in the settingsDict file.

- Change the CAD file for the hull but keep all conditions (Lpp, Fn etc.) the same in order to compare different hulls.
- Modify the propeller model and keep everything else constant to achieve a comparative study (as is demonstrated later in this paper.)
- Add and remove appendages to judge their effect on the performance (as is demonstrated later in this paper.)

- Modify the mesh density scale parameter (1,2,4,8 etc.) to achieve a mesh independence study.

FRAMEWORK FOR PROPELLER MODELING

One major obstacle when considering the modelling of combined ship/propeller flows is the treatment of the rotating propeller. Creating a rotating sub-domain/mesh with a detailed resolution of the propeller blades is both costly in terms of computational resources, but also complex in terms of the geometry; especially for propellers rotating very close to (or even inside) stationary features such as rudders, ducts and ESDs.

For obtaining the average flow (and sometimes even the unsteady flow) in such situations, it can be beneficial to approximate the propeller as a distributed momentum source in the main fluid domain. This allows for a single domain to be used and gives the user freedom to design the mesh as necessary to capture the flow around details in the stern geometry (such as ESDs) without concern for fitting an adjacent rotating mesh. In this case, the action of the propeller is modelled separately from the main flow solver. Apart from making the domain discretization much easier, modelling the propeller flow separately also has the added benefit that the propeller model can be tuned to the local Reynolds Number of the blades rather than the (very different) Reynolds Number of the moving ship (Molland et.al., 2011).

For the main flow domain containing the hull, ESDs, rudder etc., consider the flow field $\overline{\mathbf{u}} = (u, v, w)$ being affected by the force (or momentum source) $F_v = (F_{vx}, F_{vy}, F_{vz})$, The Navier-Stokes momentum equation states that

$$\begin{aligned}
\frac{\partial(\rho u)}{\partial t} + \nabla\cdot(\rho u\overline{\mathbf{u}}) &= -\frac{\partial p}{\partial x} + \frac{\partial \tau_{xx}}{\partial x} + \frac{\partial \tau_{yx}}{\partial y} + \frac{\partial \tau_{zx}}{\partial z} + \rho F_{vx} \\
\frac{\partial(\rho v)}{\partial t} + \nabla\cdot(\rho v\overline{\mathbf{u}}) &= -\frac{\partial p}{\partial y} + \frac{\partial \tau_{xy}}{\partial x} + \frac{\partial \tau_{yy}}{\partial y} + \frac{\partial \tau_{zy}}{\partial z} + \rho F_{vy} \\
\frac{\partial(\rho w)}{\partial t} + \nabla\cdot(\rho w\overline{\mathbf{u}}) &= -\frac{\partial p}{\partial z} + \frac{\partial \tau_{xz}}{\partial x} + \frac{\partial \tau_{yz}}{\partial y} + \frac{\partial \tau_{zz}}{\partial z} + \rho F_{vz}
\end{aligned} \tag{1}$$

This is known as "body force modeling" and the challenge is to find a model for F_v that accelerates the flow in the same manner as the rotating propeller would; while at the same time calculating what the loading on the blades would be to deduce the propeller thrust and torque. Naturally, F_v will be zero everywhere in the computational domain, except where the propeller should be.

There exists a range of different models for the distribution of F_v representative of a ship propeller. From a simple actuator disk where the force is spread over the radius according to an ideal distribution, to a full panel code for the propeller where the force is obtained from the pressure on each panel (Molland et.al., 2011). Which one to choose will depend on the level of complexity needed for a certain study and the available resources.

The selfPropFoam package, originally developed by Windén (2014) and Windén et.al. (2014), introduces a modular way to couple an arbitrary model for F_v with any of OpenFOAMs Finite Volume solvers. It includes

- A custom propeller control class that handles all coupling. This means that only a handful of lines of code needs to be added to any of OpenFOAMs solvers to make it a coupled "body force" solver.
- Modular approach to creating and selecting propeller models, RPM control schemes and field mapping. This makes comparative studies between different models and different schemes easier to conduct.
- A custom "concentric" propeller mesh class to be used when creating propeller models; featuring:
    - Automatic mapping of fields (such as force or velocity) from/to the main flow solver.
    - Functions for integration and differentiation of fields (such as force or velocity).
    - Curve fitting of any propeller geometry data to fit the given mesh resolution.
- Utilities to run the selected model decoupled for testing- or open water performance purposes.

By default, selfPropFoam comes with two propeller models pre-defined. The Yamazaki simplified propeller theory and Blade Element Momentum theory (BEMt). It also comes with a template model, a simple actuator disk, that is well commented and can be used as a basis to create new models.

PROPELLER THEORY

Blade Element Momentum theory combines axial momentum theory and 2D blade element theory; as first suggested by Burrill (1944). The default implementation of the BEMt in selfPropFoam follows the well-documented approach by (Molland et.al., 2011).

The implementation of the Yamazaki model is described here (briefly) since few descriptions in English exist in the literature. Originally developed by Yamazaki (1968), the simplified propeller theory has been further improved by Moriyama (1979) and Yamazaki (1998). It represents the propeller interacting with the fluid by distributing bound vortices of strength Γ in place of the propeller blades and a free vortex with pitch h to represent the trailing wake. $\Gamma(r,\theta)$ is discretely distributed over a concentric grid $[r\,,\,\theta]$ centered on the propeller centroid. This achieves a hybrid theory, merging aspects of lifting line and lifting surface theory.

$\Gamma(r,\theta)$ and $h(r)$ are found iteratively from a combination of 2D airfoil theory, potential flow, the propeller inflow and the interaction of the free vortex with the blades at the propeller plane.

The velocity disturbance caused by the propeller is described by the velocity potential $\varphi_{Pf\infty}$ which in turn is based on the Greens function G_P. The Greens function describes the cumulative influence of $\Gamma(r'\,,\,\theta')$ from each grid-panel on the total disturbance at $[r\,,\,\theta]$. Here, r_0 is the propeller radius and r_h is the hub radius.

$$\varphi_{Pf\infty} = \frac{1}{4\pi}\int_{r_h}^{r_0} r'dr' \int_0^{2\pi} \Gamma(r',\theta')G_p d\theta' \tag{2}$$

$$G_p = \frac{r'}{h(r')R_{20}} - \frac{r\,sin(\theta'-\theta)}{r'^2+r^2-2r'rcos(\theta'-\theta)}\left(1+\frac{x}{R_{20}}\right) \tag{3}$$

The disturbance is inversely proportional to the distance R_{20} from the source where

$$R_{20} = \sqrt{x^2+r'^2+r^2-2r'rcos(\theta'-\theta)} \tag{4}$$

and x is the axial coordinate; with $x=0$ at the propeller centroid.

With the disturbed velocity, the wake velocity, the propeller geometry, the strength of the bound vortices and the pitch of the free vortex known, the propeller boundary condition (velocity at the blades) can be formulated as

$$\begin{aligned}\left(\frac{2\sqrt{r^2+a(r)^2}}{Zk_1c(r)} + \frac{r^2+h(r)^2}{2rh(r)\kappa_N(r)}\right)\Gamma(r,\theta) + \left[\frac{\partial\varphi_{Pf\infty}}{\partial x}\right]_P - \frac{h(r)}{r}\left[\frac{\partial\varphi_{Pf\infty}}{r\partial\theta}\right]_P \\ = \frac{a(r)}{r}(\Omega r + [u_{P\theta}(r,\theta)]_P) - [u_{Px}(r,\theta)]_P\end{aligned} \tag{5}$$

Here, $a(r)$ and $c(r)$ describe the blade pitch and chord distributions, Z is the number of blades and Ω is the propeller rotation rate. The $[\ ]_P$-index indicates values on the propeller disk ($x=0$) and $u_{Px}(r\,,\,\theta)$, $u_{P\theta}(r\,,\,\theta)$ are the undisturbed axial and tangential velocity distributions in the (nominal) wake. $\kappa_N(r)$ is the Prandtl tip correction factor and k_l is an empirical correction for lift slope, defined as

$$\kappa_N(r) = \frac{2}{\pi} cos^{-1}\left(e^{\left(-z\left(1-\frac{r}{r_0}\right)\frac{\sqrt{r_0{}^2+h(r)^2}}{h(r)}\right)} \right) \tag{6}$$

$$k_1 = \left[1.07 - 1.05\frac{c(r)}{r_0} + 0.375\left(\frac{c(r)}{r_0}\right)^2\right]_{r=0.7r_0} \tag{7}$$

Since $u_{Px}(r, \theta)$ and $u_{P\theta}(r, \theta)$ are defined as undisturbed, in order to achieve successful coupling with a RANS solver, it is necessary to know what the total induced velocities of the propeller are. These can then be deducted from RANS velocities appearing on the propeller plane to yield the undisturbed wake. The total velocities are also used to determine the momentum imparted on the fluid, from which the thrust and torque can be deduced. The induced velocities in the axial and tangential directions are defined as

$$V_{Px} = \frac{r\Gamma(r,\theta)}{2h(r)\kappa_N(r)} + \left[\frac{\partial\varphi_{Pf\infty}}{\partial x}\right]_P \quad , \quad \overline{V_{Px}} = \int_0^{2\pi} V_{Px}\, d\theta \tag{8}$$

$$V_{P\theta} = -\frac{\Gamma(r,\theta)}{2\kappa_N(r)} + \left[\frac{\partial\varphi_{Pf\infty}}{r\partial\theta}\right]_P \quad , \quad \overline{V_{P\theta}} = \int_0^{2\pi} V_{P\theta}\, d\theta \tag{9}$$

The induced velocity in the radial direction, and the radial force are both zero by definition. From these velocities, the pitch of the free vortex can be calculated as

$$h(r) = k_2 r \frac{\overline{V_{Px}}}{\overline{V_{P\theta}}} \tag{10}$$

where

$$k_2 = 1 + 0.625\left(\frac{c_{max}}{r_0} - 0.84\right) \tag{11}$$

The total thrust T and total torque Q can be calculated by integrating the induced momentum as

$$T = -F_{Px} = -\rho \int_{r_h}^{r_0} dr \int_0^{2\pi} \frac{d^2F_{px}}{rdrd\theta} rd\theta + \frac{dF_{fx}}{dr} \tag{12}$$

$$Q = M_{Px} = \rho \int_{r_h}^{r_0} r\, dr \int_0^{2\pi} \frac{d^2F_{p\theta}}{rdrd\theta} rd\theta + \frac{dF_{f\theta}}{dr} \tag{13}$$

where the force components are

$$\frac{d^2F_{px}}{rdrd\theta} = \Gamma(r,\theta)V_{P\theta} \tag{14}$$

$$\frac{d^2F_{p\theta}}{rdrd\theta} = \Gamma(r,\theta)V_{Px} \tag{15}$$

and

$$\frac{dF_{fx}}{dr} = -\frac{1}{2}C_{PD}Zc(r)\sqrt{1+\frac{h(r)^2}{r^2}}\overline{V_{Px}}\,\overline{V_{P\theta}} \tag{16}$$

$$\frac{dF_{f\theta}}{dr} = -\frac{1}{2}C_{PD}Zc(r)\sqrt{1+\frac{h(r)^2}{r^2}}\overline{V_{P\theta}}^2 \tag{17}$$

Applying the forces calculated from Eqns. (14)-(17) as momentum sources in the main flow solver (through Eqn. (1)) and using the calculated thrust and torque from Eqns. (12) and (13) to control the rotation rate gives a model that is able to predict the performance of a certain hull/propeller combination.

COMPARATIVE STUDY OF TWO PROPELLER MODELS FOR ESD PERFORMANCE

The state of the art of ship performance CFD can be demonstrated from the results of a recurring set of workshops. These are designed to test the abilities of world-leading institutions and CFD codes to predict the performance of benchmark hulls and propellers. The latest workshop, held in Tokyo in 2015, was the first to feature a ship with an ESD as a benchmark case (Hino et.al, 2020). The Japan Bulk Carrier (JBC) is representative of a modern Capesize bulk carrier fitted with an ESD in the form of a duct in front of the propeller. This ESD was designed to improve the propeller inflow and obtain a higher level of efficiency by lowering the effective wake fraction.

Experimental data (measured in a towing tank) for the performance of the JBC, with and without the ESD, was made available as part of the Tokyo 2015 workshop and was recently published together with results from all participants by Hino et.al (2020). The JBC benchmark case is used here to demonstrate the capabilities of SHORTCUt and the selfPropFoam framework. This calculation is included as a tutorial when downloading the SHORTCUt package. Calculation settings for a Container Ship and a Tanker from the same workshop series are also included as tutorials.

The particulars of the JBC are shown in Table 1. An overview of the hull shape, including the ESD is shown in Figure 1. No rudder was present in the benchmark case and in the towing tank experiments, so it has been excluded in this study.

Table 1: Properties of the JBC benchmark hull

Property	Model Scale	Full Scale
Lpp	7m	280m
Speed (V_{ship})	Fn=0.142 (1.179 m/s)	Fn=0.142 (14.5 kn)
Wetted surface area (*S*)	12.27 m^2	19633.90m^2
	Propeller	
Diameter ($2r_0$)	0.203m	8m
No. blades (*Z*)	5	5
Rotation direction	Clockwise	Clockwise

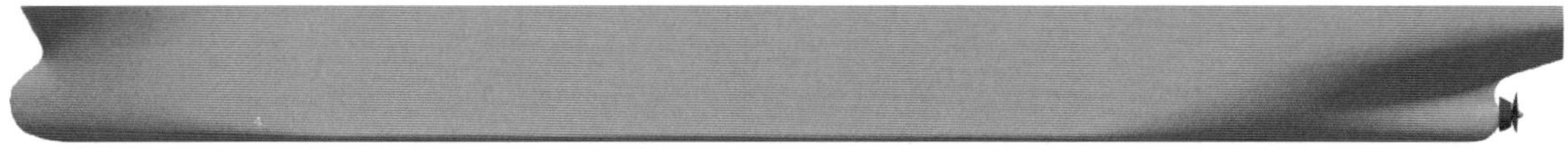

Figure 1: Overview of the JBC hull. No rudder is present in this study to conform to the conditions in the towing tank experiment.

The flow around the JBC, as well as relevant coefficients used to judge the efficiency of the ESD are computed using both the BEMt and Yamazaki models. The comparison between the two propeller models is made while keeping everything else constant; clearly illustrating the benefits of the modular approach of selfPropFoam. Both propeller model calculations are done on the same concentric mesh with 10 radial and 36 circumferential panels. The discretization of the propeller model and the surface mesh in the stern area are shown in Figure 2.

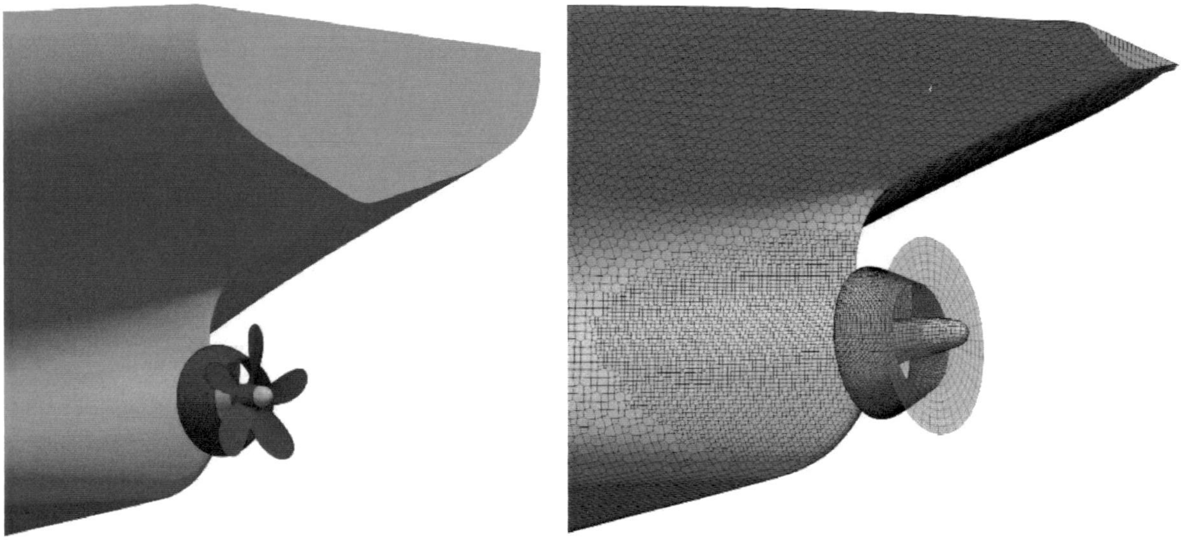

Figure 2: Detail of JBC stern area. Towing tank model design (left) and discretization, including propeller model discretization (right).

For the purposes of this study, a standard laptop computer was used for the calculations. This illustrates the accessibility for users with limited resources to these types of calculations; made possible in part due to the use of body force modelling. The specifications of the computational resources used are shown in Table 2. The quoted times for the self-propulsion test include post-RANS analysis of propulsion coefficients using an open water calculation. Generally, this takes less than one minute of computational time.

The times for the self-propulsion test in Table 2 are quoted for the Yamazaki model. Small differences can be seen in the computational time when using the BEMt model. These are in the range of a few minutes for a >10h computation. The mesh and the resistance test are shared between the Yamazaki and BEMt calculations (i.e. only generated and computed once.)

Table 2: Computational resources and specifications of computational effort.

Computational resources				
Machine	Dell XPS13 Ultrabook Laptop			
System	Ubuntu 18.04.4 LTS			
CPU	Intel Core i7-6560U CPU @ 2.20GHz × 3			
Memory	16 GB			
Case specifications				
	Without ESD		With ESD	
Cell count	7.670.762		7.867.613	
y_1^+	~1		~1	
Turbulence model	kω-SST		kω-SST	
Pressure-velocity coupling	SIMPLE		SIMPLE	
Free surface treatment	Single-phase, symmetry plane		Single-phase, symmetry plane	
Computational effort				
	Without ESD		With ESD	
	RANS iterations	Clock time	RANS iterations	Clock time
Time to create the mesh	N/A	1h 7min	N/A	1h 11min
Total setup time (including meshing)	N/A	1h 29min	N/A	1h 32min
Time to perform resistance test	2000	12h 40min	2000	12h 51min
Time to perform self-propulsion test	1500	10h 8min	1500	10h 21min
Total time for ship performance analysis	**3500**	**24h 17min**	**3500**	**24h 44min**

RESULTS

Comparisons are made here both to the results of Experimental Fluid Dynamics (EFD) made available by the National Maritime Research Institute (NMRI) for the JBC, but also with results from other CFD-based methodologies that were submitted to the Tokyo 2015 workshop. This includes numerous metrics for comparison. To limit the scope of this paper, only the wake fraction *1-w*, the thrust deduction factor *1-t*, the total estimated powering requirement for the full-scale ship P_s and contours of the wake velocities behind the propeller are discussed here.

The thrust deduction factor *1-t* is defined as

$$1 - t = \frac{T + R_{tow} - R_{prop}}{T} \tag{18}$$

Where R_{tow} is the resistance during the resistance test and R_{prop} is the resistance during the self-propulsion test. The wake fraction *1-w* is defined as

$$1 - w = \frac{V_{prop}}{V_{ship}} = \frac{J_i r_0 \Omega}{\pi V_{ship}} \tag{19}$$

where the equivalent advance ratio J_i is found by thrust identity from the open water results (for each model) in Figure 3. Finally, the overall propulsive efficiency η and the powering requirement for the ship P_s are defined as

$$\eta = \frac{1 - t}{1 - w} \eta_0 \eta_r \tag{20}$$

$$P_s = \frac{P_e}{\eta} = \frac{R_{tow} V_{ship}}{\eta} \tag{21}$$

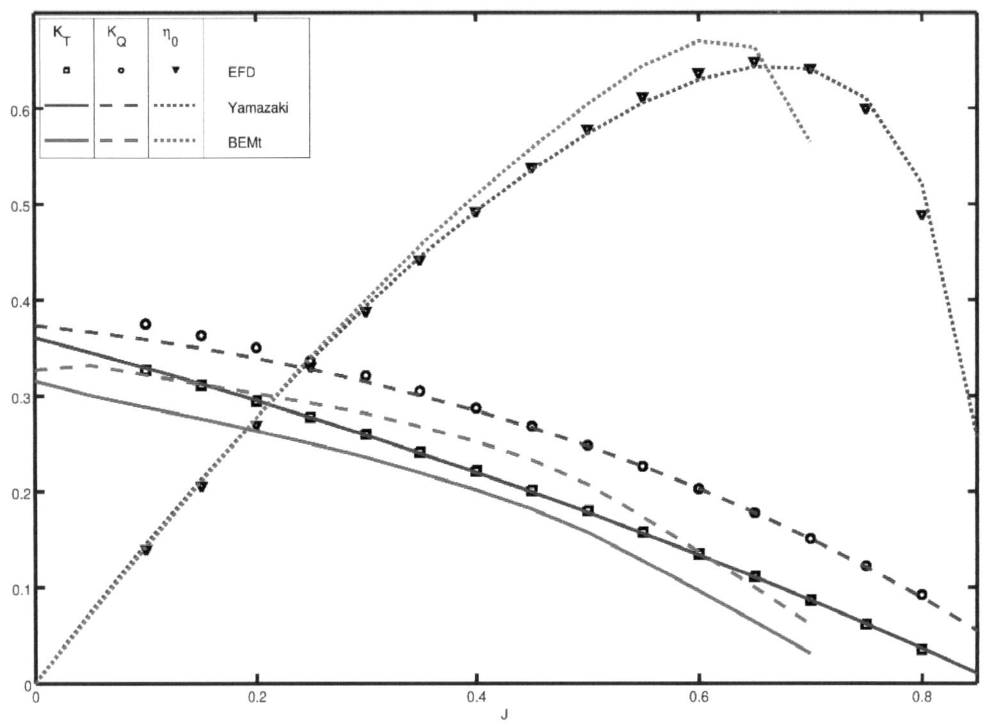

Figure 3: Predicted open water performance of the JBC propeller, using the BEMt and Yamazaki models, compared to experiments.

Figure 4, Figure 5 and Figure 6 show the predicted values of *1-t*, *1-w* and P_s respectively. The results for both the BEMt and Yamazaki models are included, together with the towing tank results. The effect of the ESD can be seen in the significantly reduced values of *1-w* and a slight reduction in the powering requirement P_s. Also included in these figures are the results submitted by other institutions to the Tokyo 2015 workshop. These are identified by institution name and the main solver used. The CFD results are also separated based on the type of propeller modeling technique used. Submissions that used a full rotating propeller simulation are marked grey, while the ones using a body force approach are marked white.

All results are ranked (from left to right) according to how close they were to the experimentally predicted value for each parameter. The results of the current study are highlighted in red and the experimental value is shown with a dashed line.

Contours of the flow in the global *x,y,z*-directions at the Aft Perpendicular (AP) are shown in Figure 7, Figure 8 and Figure 9 respectively for the case with no ESD present; and in Figure 10, Figure 11 and Figure 12 for the case with the ESD present.

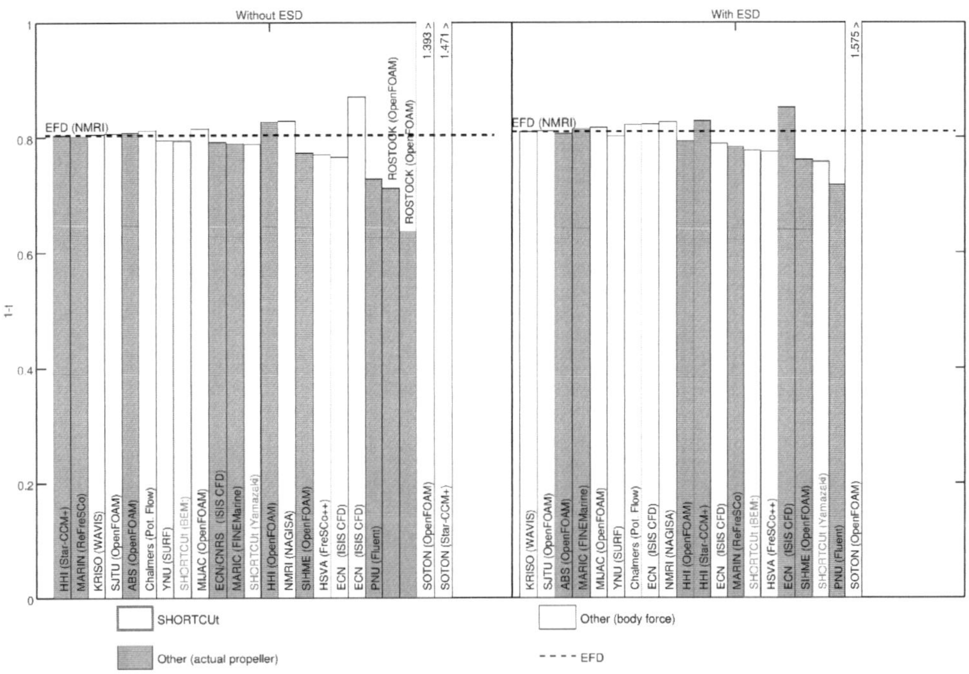

Figure 4: Predicted thrust deduction factor (*1-t*) of the JBC, with and without the ESD. Results of the present study is in white with a red outline. These are compared to submissions to the Tokyo 2015 workshop. Other studies using body forces are shown in white and other studies simulating the full rotating propeller in dark grey. Results are ranked from left to right in terms of error compared to the experiment (dashed line).

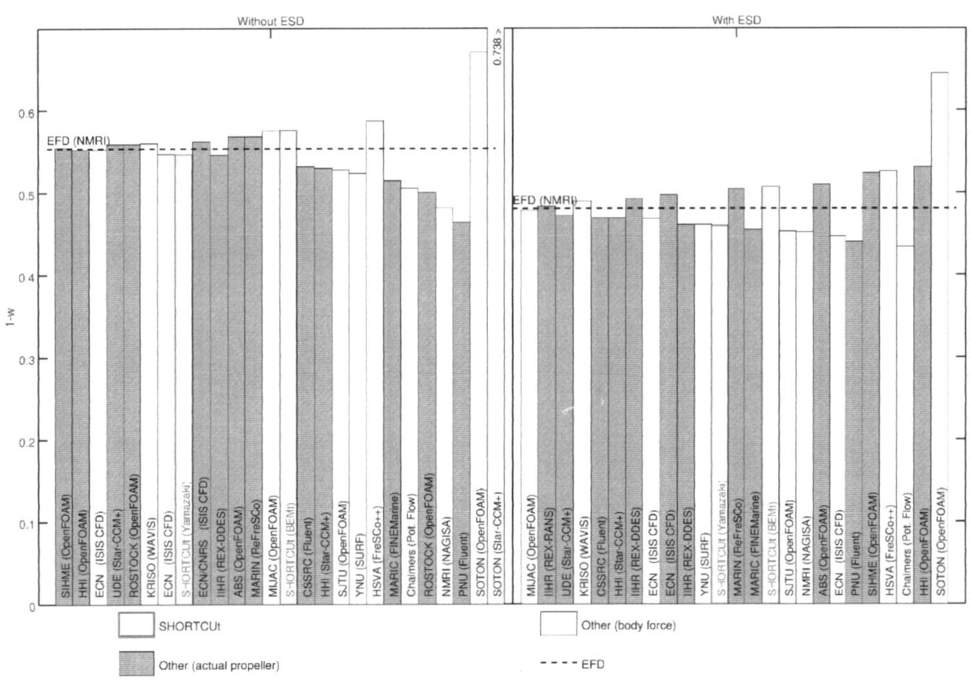

Figure 5: Predicted wake fraction (*1-w*) of the JBC, with and without the ESD. Results of the present study is in white with a red outline. These are compared to submissions to the Tokyo 2015 workshop. Other studies using body forces are shown in white and other studies simulating the full rotating propeller in dark grey. Results are ranked from left to right in terms of error compared to the experiment (dashed line).

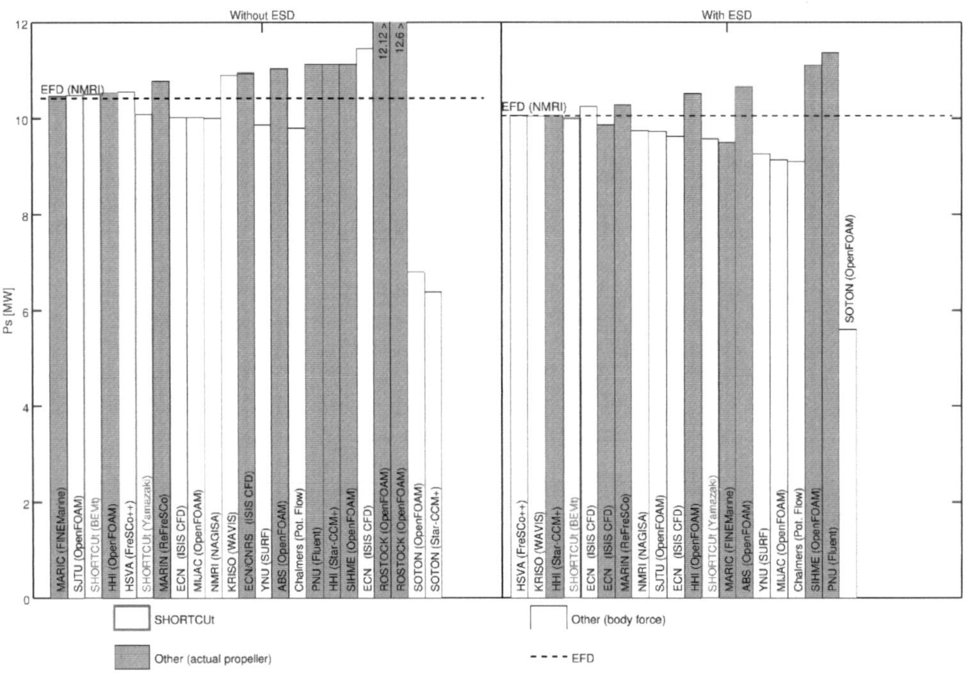

Figure 6: Predicted powering requirement of the full-scale JBC, with and without the ESD. Results of the present study is in white with a red outline. These are compared to submissions to the Tokyo 2015 workshop. Other studies using body forces are shown in white and other studies simulating the full rotating propeller in dark grey. Results are ranked from left to right in terms of error compared to the experiment (dashed line).

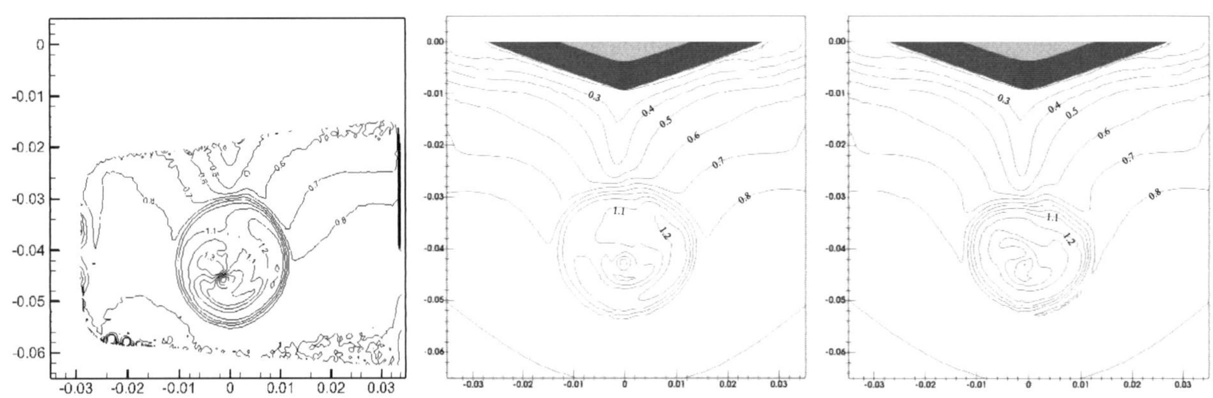

(a) (b) (c)

Figure 7: Axial velocity *u* at AP without ESD. Experimental data (a), Yamazaki model (b) and BEMt model (c).

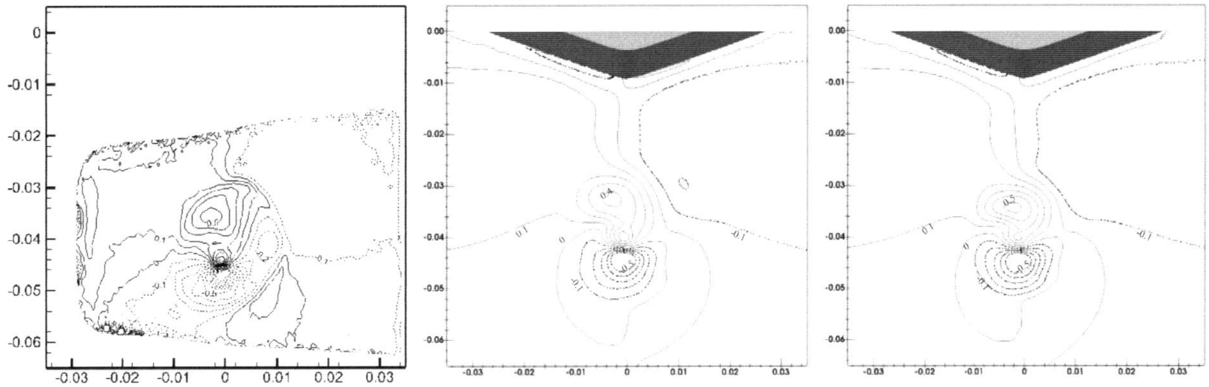

(a) (b) (c)

Figure 8: Tangential velocity *v* at AP without ESD. Experimental data (a), Yamazaki model (b) and BEMt model (c).

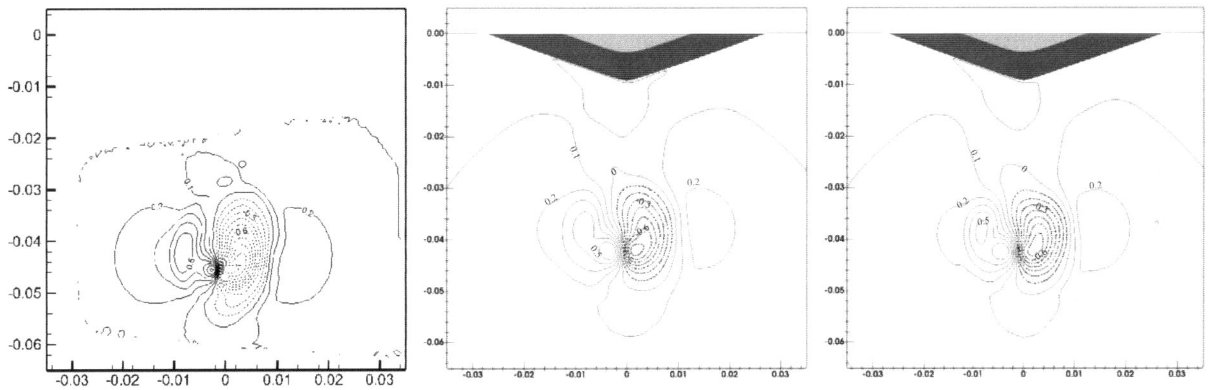

(a) (b) (c)

Figure 9: Tangential velocity *w* at AP without ESD. Experimental data (a), Yamazaki model (b) and BEMt model (c).

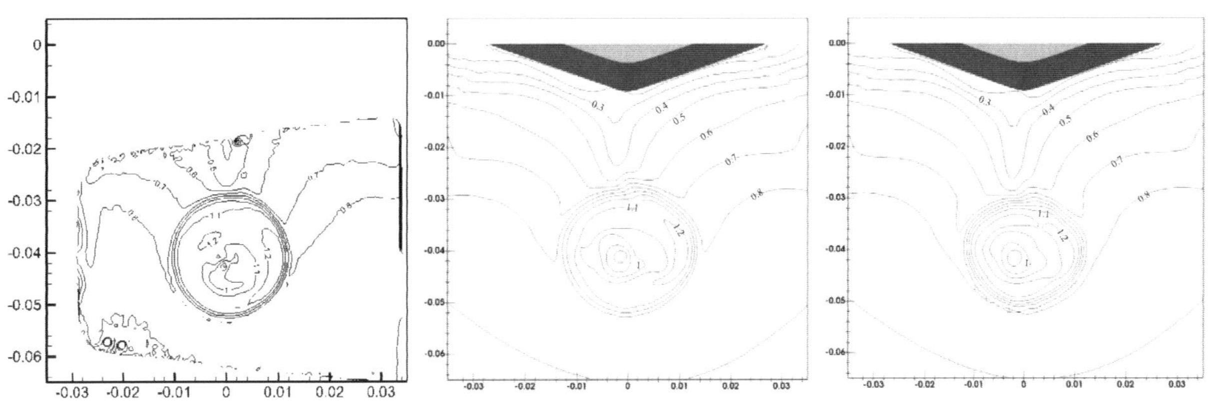

(a) (b) (c)

Figure 10: Axial velocity *u* at AP with ESD. Experimental data (a), Yamazaki model (b) and BEMt model (c).

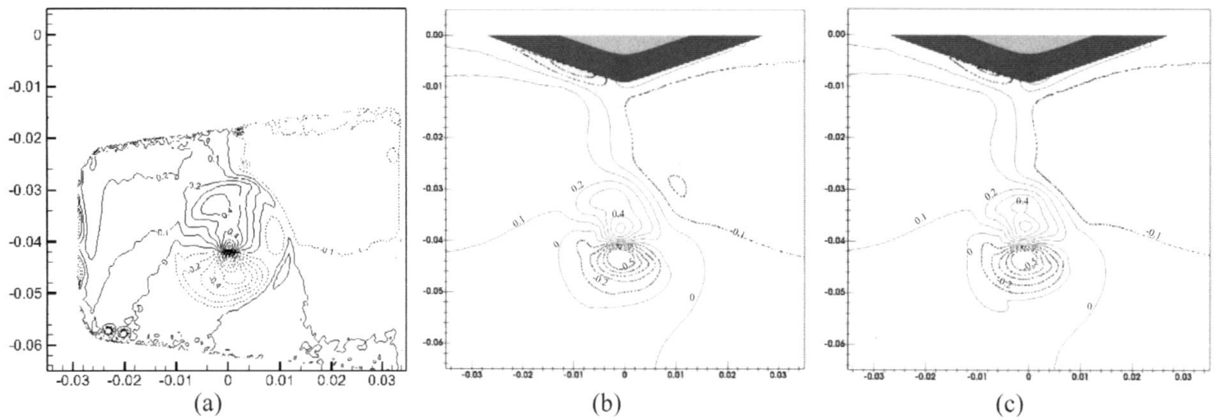

Figure 11: Tangential velocity *v* at AP with ESD. Experimental data (a), Yamazaki model (b) and BEMt model (c).

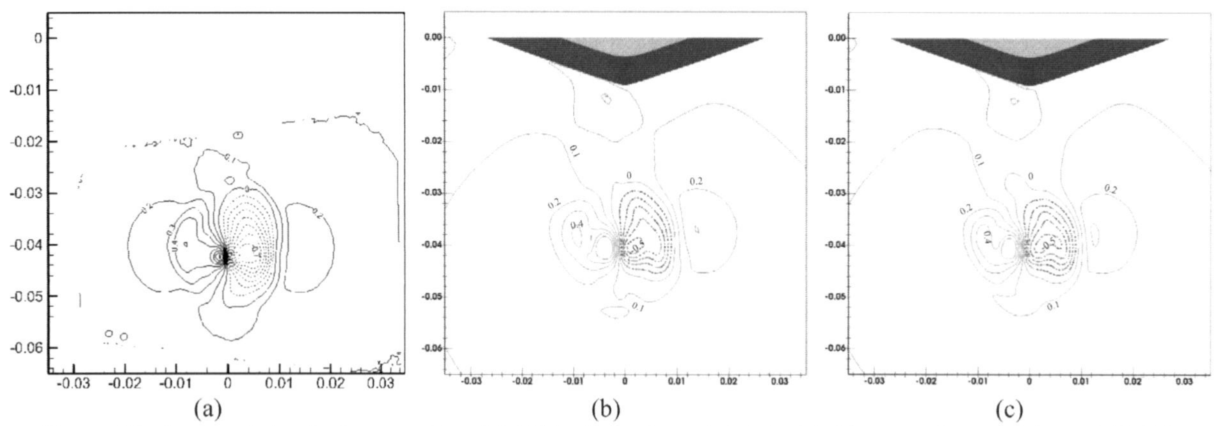

Figure 12: Tangential velocity *w* at AP with ESD. Experimental data (a), Yamazaki model (b) and BEMt model (c).

CONCLUSIONS

The results of this study are demonstrated to be in line with the state of the art of marine CFD. Prediction of all considered factors rank as average or above average compared to other leading institutions and solvers. There is no clear trend to suggest that methods that use a body force formulation are less accurate at predicting the powering performance of the JBC than the more time-consuming actual propeller simulations.

This means that a high level of accuracy at predicting the performance of the ESD is not exclusive to large institutions and commercial solvers. The results presented in this study are based on a calculation setup that is free to download and can be run on a standard laptop in around 24h. It should be noted that, to conduct this simulation in full scale, while only requiring minimal adjustments to the setup in SHORTCUt, would involve a considerable increase in the computational effort required.

The Yamazaki model is superior when it comes to capturing the local flow regime near the propeller. The BEMt suffers from certain assumptions in the underlying momentum theory that make the propeller-induced velocities more difficult to deduce. Furthermore, the BEMt model does not deal well with the effects of the tangential wake; an effect that can be seen in the distorted high-velocity area in Figure 7 and Figure 10. However, the average performance estimates delivered by the BEMt on par with (and sometimes better than) the ones obtained from the Yamazaki method. The resistance was slightly under-predicted in these simulations, so it is possible that an under-prediction of the efficiency by the BEMt has offset this to create an overall better prediction of the powering requirement.

The mesh around the stern area was relatively coarse in this study. This was partly due to the limited computational resources. An earlier study using the selfPropFoam framework, the Yamazaki model and a finer mesh was presented at the Tokyo workshop by Windén & Eguchi (2015) with similar results for the overall performance but better predictions of the local flow. This study can be seen in Figure 4, Figure 5 and Figure 6 marked as "MIJAC (OpenFOAM)".

While the results of this study are encouraging, it should be noted that they should not be taken as a sign that the method will always give accurate results. Different setups will be suitable for different types of hulls and ESDs. Fortunately, several other benchmarking cases are available in the same workshop series as the JBC. When downloading SHORTCUt, a user has access to validated calculation setups to match many of these, including a container ship, tanker and navy destroyer. When seeking the performance of a new hull, the validated calculation setup that is closest to being representative of the ship type in question is a good starting point.

Finally, it should be noted that the purpose of this paper has not been to showcase a perfected methodology for judging the energy-saving effects of ESDs. But rather to showcase how these types of studies could be made available to a wider range of researchers, both in industry and in academia. Readers of this paper are encouraged to implement different propeller models; and to explore different options for modelling the turbulent, multi-phase flow around the hull using the many built-in and user-made schemes, models and solvers in the OpenFOAM suite. The framework for creating coupled solvers presented in this paper provides the perfect basis for such future developments.

REFERENCES

Burrill L. (1944), "Calculation of marine propeller performance characteristics", Transactions of NECIES **60**

Carson, I. (2012), 'Slow steaming uphill', The Economist, August 1st 2012.

Herbsleb, J., Kästner, C. and Bogart, C., (2015), 'Intelligently transparent software ecosystems', *IEEE Software*, ***33***(1), pp.89-96.

Hesketh, R. and Wittels, J. (2020), 'What's the Green Fuel of the Future for Shipping?', The Washington Post, September 1st, 2020

Hino T., Stern F., Larsson L., Visonneau M., Hirata N., Kim J. (Eds.), (2020), 'Numerical Ship Hydrodynamics: An Assessment of the Tokyo 2015 Workshop', Springer, ISBN: 978-3-030-47571-0

Hiroi, T., Windén, B., Kleinwächter A., Ebert E., Fujisawa J., Kamiirisa H., Damaschke N., Kawakita C. (2019), Full-scale on-board measurements of wake velocity profiles, under-water noise and propeller induced pressure fluctuations, JSNAOE Autumn Meeting 2019, November 21-22.

IMO (2011), MARPOL Annex VI energy efficiency amendments, Technical report. Resolution MEPC.203(62).

ITTC (1999), 'Final report of the specialist committee on unconventional propulsors', 22nd International Towing Tank Conference, Seoul Korea and Shanghai, China.

ITTC, (2017a), 'Final Report and Recommendations to the 28th ITTC of the Specialist Committee on Energy Saving Methods'

ITTC, (2017b), Recommended Procedures and Guidelines 7.5-02-03-01.4: 1978 ITTC Performance Prediction Method

Kim, M.C., Shin, Y.J, Lee, W.J and Lee, J.H. (2017), 'Study on Extrapolation Method for Self-Propulsion Test with Pre-Swirl Device', Fifth International Symposium on Marine Propulsion smp'17, Espoo, Finland.

Marsan, J., Templier, M., Marois, P., Adams, B., Carillo, K. and Mopenza, G.L., (2018), 'Toward solving social and technical problems in open source software ecosystems: using cause-and-effect analysis to disentangle the causes of complex problems'. *IEEE Software*, ***36***(1), pp.34-41.

Molland, A., Turnock, S. and Hudson, D. (2011), 'Ship Resistance and Propulsion: Practical Estimation of Ship Propulsive Power', Cambridge University Press.

Moriyama F., (1979), 'On an Approximate Numerical Method for Estimating the Performance of Marine Propellers (プロペラ性能の近似計算法について)', Papers of Ship Research Institute **16**(6), p361-376

OpenFOAM Foundation, (2019) 'OpenFOAM version 7', URL: https://openfoam.org/download/7-source/

Sakamoto, N., Kobayashi, H., Ohashi, K., Kawanami, Y. Windén, B., Kamiirisa, H. (2020), An Overset RaNS Prediction and Validation of Full Scale Stern Wake for 1,600TEU Container Ship and 63,000 DWT Bulk Carrier with an ESD, Applied Ocean Research **105**.

SHORTCUt CFD (2021), 'SHORTCUt v.0.11.4' URL: https://www.shortcutcfd.com/downloads, Accessed on Feb 10th 2021.

Visonneau, M., Deng, G.B., Guilmineau, E., Queutey, P. and Wackers, J., (2016), 'Local and global assessment of the flow around the Japan bulk carrier with and without energy saving devices at model and full scale'. In *31st ONR Symposium on Naval Hydrodynamics*.

Windén B., (2014), "Powering Performance of a Self Propelled Ship in Waves", Ph.D. Thesis, University of Southampton

Windén, B., Turnock, S. and Hudson, D. (2014), 'A RANS modelling approach for predicting powering performance of ships in waves', International Journal of Naval Architecture & Ocean Engineering **6**(2)

Windén, B. and Eguchi, T., (2015), Verification and Validation of a Coupled Solver for Predicting Self Propulsion of the JBC Bulk Carrier, Tokyo 2015 A Workshop on CFD in Ship Hydrodynamics, Tokyo, Japan

Yamazaki R., (1968), 'On the Propulsion Theory of ships on Still Water -Introduction-', Memoires of the Faculty of Engineering, Kyushu University, **27**(4), p187-220

Yamazaki R., (1998), 'Deduction of the Simplified Propeller Theory (簡易プロペラ理論について)', Transactions of the West-Japan Society of Naval Architects (西部造船会会報), **95** p251-271

SNAME
THE INTERNATIONAL COMMUNITY
FOR MARITIME AND OCEAN PROFESSIONALS

Sustainable Path for the Oil and Gas Industry

MADINA INTYKBAYEVA
AFFILIATION (MCDERMOTT)

Proceedings of the 26th Offshore Symposium, April 6-7, 2021, Houston, Texas
Texas Section of the Society of Naval Architects and Marine Engineers

ABSTRACT

This paper focuses in the discussion of three main transformations the Oil and Gas Industry will need to continue developing post-pandemic scenarios. Sustainability, Digitalization, Cultural Innovation and Branding need to continue its parallel development for the industry to keep the leadership positions in the energy sector. The goal of this paper to show the interdependence between these three transformations and how EPCI companies need to continue adapting them to succeed.

Introduction

The Oil and Gas Industry is fundamental and crucial to today's world as our energy, transportation and way of living depend on many primary and secondary sources that are produced by oil and its sub products. However, the industry must go through three fundamental changes to survive, adapt and maintain the work successfully in the upcoming years.

Recently, the oil and gas has faced various challenges, which to this day have now a front seat within states and international organizations. A drop in oil demand due to the overproduction, increases developments in renewables solutions, and concerns of the young generation about climate change and environmental problems are key factors, which the industry continues to face.

This paper will discuss the main three major shifts (Sustainability, Digitalization, Cultural Innovation and Branding) – transformations that need to happen within the industry and explain why it is important to go through them simultaneously.

World economic crises, pandemics, and a shift towards cleaner solutions have played a big role in this world demand drop. The COVID-19 pandemic has helped accelerate what is already shaping up to be one of the industry's most transformative moments.

A fundamental change of application, processes, know-how, and culture needs to occur, to maintain the industry in an ever-evolving globalized world. New businesses related to the energy transition and renewables will continue to emerge. According to a Morgan Stanley survey, 85% of individual investors are interested in sustainable investing and as many as 95% of millennials polled in the survey expressed interest in investing in sustainable companies. Major companies like Shell are linking their credit facilities with their emissions-reduction targets, and they are reducing their cost of capital as well. Shifting and taking a lead in the energy transition will help oil and gas companies to be on demand for investors and young generation employees.

Hydrogen, ammonia, methanol, new plastics, and carbon capture, utilization, and storage (CCUS) could all be potential areas for the oil and gas industry to achieve leadership roles. The key for the industry is to incorporate a digital and social transformation in parallel with sustainability to achieve leadership role. Companies cannot continue to operate as a business as usual. Due to COVID-19, many projects went virtual. For example, the distant projects started to use virtual and augmented realities to overcome traveling restrictions. All project meetings went online, and with reduction of travel shifted the way we socialize now. For example, digital communication solutions that are provided by Microsoft such as Teams, Viva, and collaboration with Spatial on holographic meetings and HoloLens will become new normal. The use of drones and automated machines will de-man the plants and fabrication sites. New emerging technologies such as Internet of things (IoT), Big Data Analysis, Artificial Intelligence (AI) and Machine Learning (ML) can and will transform operations, enhance agility and help in strategic decision making, and build new business models and predictive models.

Global Trade's studies show that critical factors such as generational gaps and hiring processes have affected digital innovation. If a company is known for extensive hours, poor culture and lack of technology, millennials, and Gen Z will move to another industry or competitor. On average less than 3% of the yearly gross profit of an oil and gas company is being reinvested in research and development, while companies such as Google invest up to 20% of their profits in Research and Development (R&D).

Artificial Intelligence and Machine Learning can capture and decrease significantly errors during the engineering design which is crucial for Engineering, Procurement, Construction and Installation (EPCI) companies that provide service to operators. By analyzing pattern recognition and hiring people whose main skills are data analysis and software engineering, can help significantly to improve productivity and efficiency by moving the knowledge into the digital world and developing user interface for upcoming projects and to maintain the processes.

The loss of revenue is no longer being tolerated within oil and gas prices being low, and when combined with the drive for sustainability – increasing efficiency and returns for investment on data are driving change. With prices fluctuating more than ever, digital disruptors must become the new norm. Digitalization is a key part of reducing cost, efficiency, and productivity, however it takes longer time for oil and gas industry due to the generational gap and lack of awareness and shift in culture. The oil and gas industry always valued experience over innovation due to safety aspects, however, it created a narrow path for the younger generation to raise their voices and shift the business perspectives.

It is a perfect time to attract new talent with a different set of skills to accelerate digital transformation. Companies also need to maintain employees who are highly adaptive to digitalization because those are at high risk of migrating to other industries (e.g., technology and consulting firms, digital solution providers, etc.) where the prospects of career growth seem to be brighter and safer. Organizations need to create space for new skills that will enhance work practices.

With the pandemic and the current health crisis, the social transformation has become visible to all. With massive layoffs across the industries, pay inequality, social dilemmas have emerged affecting the industry's reputation and employee's reliability in it. To improve the situation new marketing and branding strategies should be developed to attract younger generations and investors who currently obtain news from social media and prioritize sustainability.

Methodology

This paper is based on a review of McDermott's study work on facilities of the future and decarbonizing EPCI projects, literature review of digitalization and carbon footprint, work experience in the industry, and a review of social media profiles of major energy companies.

1st Transformation: Sustainability

After the Paris Agreement in 2015 was signed, societal pressure to reduce carbon footprint, risks, and impacts of climate change begun to grow for all industries. The oil and gas industry is getting more pressure to reduce emissions and meet the requirements for receiving a "social license to operate." Many countries and government regulatory agencies are shifting their strategies to address this continuous pressure. For example, at the G-20 summit, the King of Saudi Arabia, Salman bin Abdel Aziz Al Saud, said that his country has big plans for solar and wind generation, which will claim 50% of the kingdom's energy by 2030.
Another major shift is that Joseph R. Biden Jr. got elected as a USA president and now brings with him the largest team ever assembled inside the White House to tackle global warming. "Since the U.S. has one of the biggest economies in the world and has contributed the most to climate change, it is incredibly important that the U.S. return to the Paris agreement," said Cornell University climate scientist Natalie Mahowald, a lead author of the 2018 U.N. report on climate change. Biden has re-entered Paris Agreement as well as plans to extend beyond Paris, including a $2 trillion economic plan to invest in a transition from fossil fuels to clean energy, cut carbon emissions from electric power to zero by 2035 and reach net-zero emissions by 2050. Those investments and changes of government strategy will help to

introduce, support, and leverage studies and research in the USA by getting funding from the government to complete several studies such as Hydrogen Storage and Carbon Capture Technologies.

In Europe, the energy transition to renewable energy took place gradually, primarily due to the expansion of the market and a decrease in coal generation. Oil and gas production in Europe fell for natural reasons, the import of hydrocarbons grew relative to the changing infrastructure, considering the development of the renewable energy sector.

In the case of the United States and Saudi Arabia, the conversation is about artificially reducing domestic consumption of hydrocarbons. But this does not mean that those countries will stop the production of oil and gas because they can profit from exporting it. This will lead to reducing investments in the oil and gas sector in favor of renewable energy and other sectors of the economy. If shortly the consumption of oil and gas in the country begins to fall, and production remains at the same level, then this will be akin to a serious increase in world production.

Following the Paris Agreement, oil and gas giants also started to announce net-zero goals such as BP, Shell, Eni, Equinor, Repsol, TOTAL. As well as others announce Oil and Gas Climate Initiative (OGCI) emissions intensity targets: Chevron, ExxonMobil, BR, Saudi Aramco. Sustainability is becoming a new normal for operators, and EPCI companies need to lead the effort and be able to meet the new requirements and demand set by operators.

Renewables and the traditional oil and gas industry can co-exist and benefit from each other. According to Deloitte, which recently issued its 2021 industry outlook after surveying a broad array of U.S. executives. For natural gas, about 58% said they believe the commodity remains key as operators position their portfolios for lower-carbon resources. Another 42%, however, see a reduced or varied role as renewables markets strengthen and industries regionalize.

EPCI companies such as McDermott, Technip, Flour, KBR, Chiyoda, Saipem and Bechtel need to lead the research on decarbonizing facilities through collaboration with other companies/startups, receiving grants and support from the government, as well as considering the acquisition of companies that are becoming experts in renewables. Service companies need to adapt to the new demand faster than operators and take a lead to be able to provide those services very soon. EPCI companies will not be able to easily enter the renewables market, and it would be extremely hard to restructure the processes unless acquisitions and restructuring will take place.

So why should the oil and gas go through this transformation and lead the energy transition?
Because oil and gas industry have the assets, knowledge, and capital to become a leader in the energy transition. Demand for crude oil and natural gas will remain in place, however the current impact the industry is having in the environment and its association to climate change will not be tolerated anymore. EPCI companies with operators' support need to adapt by investing in research and studies, collaborating with stakeholders, or developing and implementing solutions to reduce emissions across oil and gas infrastructure. Service companies can develop low carbon strategies and solutions for projects, including liquefied natural gas (LNG), Offshore Platforms, CO2 capture technology, hydrogen storage, and smart modularization by utilizing and transforming, and adjusting the knowledge and expertise towards emissions reduction.

Considerable research and collaborative efforts exist in building construction, where the global community has recognized the importance of decarbonization of the building value chain to achieve emissions goals. However, there is little evidence of research and collaboration on the construction carbon footprint of large-scale energy infrastructure projects. Even in building studies, gaps exist in lifecycle assessments covering the different phases of the building lifecycle, particularly in construction. The raw materials are similar in buildings and industrial construction, but the equipment, execution, and processes involved differ in complexity and scope. As a result, the oil and gas industry can learn from advanced sectors like buildings, technology, and renewable energy but must come together to address the specific challenges of CAPEX emissions in oil and gas projects, particularly from a value chain (client to supply chain) standpoint.

One of the hardest but major shifts needs to happen in the offshore industry. According to EPCI companies' emissions data, vessels have the largest footprint. For example, McDermott has a fleet of derrick barges, construction and pipelay, and support vessels capable of transporting and installing complex oil and gas structures such as jackets, topsides, and subsea structures. The fleet can install pipelines in S-lay, reel-lay, and flex-lay from shallow to deep water. Global deep-water fleet is modern; however, those assets are powered by thrusters to move on their powers that require a lot of fuel. Shallow water fleet- maintain position with anchors and will need tugboats to move around. However, both types of vessels require power to maneuver vessels.

With the collaboration and support of the shipping industry and governments, there are several possible ways to decrease the carbon footprint of the vessels such as renewable energy and fuels, hydrogen and battery technology, and the use of LNG and Nuclear fuel sources.

One of the solutions is to use shore green power when vessels are not offshore. This will be the fastest and optimized solution to significantly reduce the vessels' carbon footprint. However, vessels are global assets, and they travel around the world and using different ports. Work vessels require large power plants but lack space and cannot be loaded down by additional weight, much like an offshore platform. Vessels are unlike the platforms; they are mobile and thus cannot rely on power from shore unless a network of "charging stations" existed in the sea or from an existing platform during work performed for that site. When vessels are alongside in port, shore services can be used for power. This may require capital investment (CAPEX) investment in vessels or port locations to connect the power supply and depends on local infrastructure availability in other locations.

Another solution is to use batteries to replace generators, running machines at low loads is inefficient and creates a high amount of emissions for relatively little power when berthed. With this solution, the generator can be shut down when the battery is charged. This will allow the number of engines in operation to be reduced and be more economical while in operation. This solution can not only improve performance, cut fuel consumption, and reduce CO2 emissions, but also will help save money on fuel in the long-term perspective. Shell aims to commercialize the technology, enabling more widespread use of batteries onboard offshore vessels. According to Shell, battery hybrid gives between roughly 12 and 30 percent fuel reduction when applied to peak shaving and spinning reserve. All that makes for a solid business case.

Hydrogen is one of the most promising alternative fuels for offshore vessels and is completely zero-emission when produced with renewable energy or from natural gas with carbon capture. "Green hydrogen can be emission-free on a full lifecycle basis," says Marie Hubatova, a shipping emissions expert at the Environmental Defense Fund. "That means from the point of where the fuel is extracted, or produced, until the point of combustion." However, the use of hydrogen as fuel currently faces several challenges, such as fuel storage and, in particular, a lack of bunkering infrastructure in ports. The supply chain for Hydrogen needs to be developed. "The problem is that right now, the availability of green hydrogen is just not there," says Xiaoli Mao, a researcher on the marine team at the International Council on Clean Transportation (ICCT). "Fuel producers need to see some legitimate demand to invest into its production, so it's sort of like a chicken/egg problem – whether the ship technology develops first, or the fuel side develops first," she says. Supplementing hydrogen with batteries can help to maintain the vessel and support the peak demand required for the Deepwater fleet. Or use green ammonia, which is another fuel that can be either combusted or used in the fuel cell and it is easier to store than hydrogen that requires not only refrigeration but cryogenic temperatures. Ammonia will take less space due to density however it is toxic for humans and the environment, as well as more expensive due to extra power requirements for conversion from hydrogen to ammonia.

Biofuels – fuels made from plant materials or animal waste – are another alternative fuel that can be used for offshore vessels. The US Department of Energy defines renewable diesel as, "a biomass-derived transportation fuel that is chemically similar to petroleum diesel and is suitable for use in conventional diesel engines." In the US, it is governed by the ASTM D975 specification. Renewable diesel can be produced from a variety of processes, including "hydrotreating, gasification, pyrolysis and other thermochemical and biochemical means, and can be made from lipids and cellulosic biomass (such as crop residues, woody biomass, and dedicated energy crops)." Renewable diesel produces significantly fewer emissions over the lifecycle of the fuel. Neste MY Renewable Diesel, for example, claims to reduce GHG emissions by up to 90% over the lifecycle of the fuel compared to fossil diesel.
The US Environmental Protection Agency has completed a thorough lifecycle assessment of biodiesel and renewable diesel lifecycle carbon footprints. Since renewable diesel can be substituted as a drop-in fuel, 100% renewable diesel reduces GHG emissions by 5-90% depending on the feedstock.
Battery technology, hydrogen and biofuels are just some areas that EPCI companies that provide services for operators should consider and investigate the possible alternative fuels that can help to reduce the carbon footprint and still be economically paid off.

Another way for EPCI companies to reduce offshore oil and gas footprint is the engineering design, fabrication, and installation of offshore platforms. Service companies can start offering and designing Net Zero Platforms that will significantly reduce carbon emissions. Based on McDermott and Schneider Electric's research, most operational emissions are related to power generation. The first step in designing a Net Zero facility is to decarbonize power. The best approach to de-carbonizing power on a compression facility is to import power from appropriate renewable sources, as well as developing a facility that has minimal carbon intensity and infrastructure. This is achieved using equipment and instrumentation that minimizes and eliminates potential leaks or venting requirements. Minimizing infrastructure should minimize embedded carbon emissions and hence support the minimization of carbon intensity. By maximizing innovation and technology to unman the facilities, use of robots and machines to reduce accommodations on platforms and run the platform remotely from shore will also continue in the overall effort to reduce carbon footprint. However, that's only possible if oil and gas companies will start investing and collaborating with tech giants to develop new Net Zero Platforms that can operate remotely.

The second step to achieving net zero is to address fabrication and construction operations. Based on EPCI companies' emissions data, fabrication yards and construction sites are the second in line to produce a significant amount of carbon emissions after offshore vessels. Power and fuel supply are the main source of emissions. However, power is particularly challenging at the fabrication and construction sites due to locations of fabrication yards that do not provide ideal conditions for renewable sources, such as local wind or hydropower generation. Some fabrication yards are based in countries with insufficient power generation and no renewable grid or with regulatory challenges to producing renewable power. The simplest option to reduce carbon footprint in fabrication is to connect to a renewable grid or establish a power purchasing agreement. Without this infrastructure, onsite small-scale solar and wind energy can be used, but further decarbonization would require advances in energy storage and renewable energy supply.

Hydrogen can also play a big role in decarbonizing facilities by utilizing a strategy of fuel cell electric vehicles, forklifts, cranes, and trucks. The movement of materials and loading and unloading can happen with unmanned guided vehicles that are hydrogen powered. Not only it is environmentally friendly, but it also can improve efficiency and reduce labor.

Another method for EPCI companies that both operate and provide services in upstream and downstream sectors is to apply the offshore technique of modularizing the onshore plants such as LNG and Petrochemicals. Modularization is a generic term applied to the project execution method of conducting an amount of construction away from the plant location site into a controlled environment to improve material, resource, and schedule efficiencies. This drives construction closer to a manufacturing environment, generally achieved by conducting pre-fabrication and sizable pre-assembly of the plant in transportable units, or modules. Modularization has moved beyond just offsite construction fabrication; innovative project teams are pursuing a true plug-and-play approach that is driving reductions in construction costs and schedule durations. It is often considered more cost-effective, and safer than building onsite, especially in a live low carbon facility with the use of technology and innovative early engineering design stage approach. If to move most of the construction work to the yard, less labor in the field will be required, and the use of controlled conditions at the yard for construction, can improve quality, reduce the GHG emissions at the yard and at site, and reduce waste. All which at the end will reduce the job site environmental impacts and onsite materials storage. This is also beneficial from the safety standpoint, through lower exposure to field hazards and through modularization will allow reduced the plot plan for future plants and facilities.

However, currently, transportation and additional use of steel during the sea fastening can cause an extra amount of carbon footprint. The reduction in the quantity of piping and concrete in modularization has to be weighed against the increase in structural steel transportation. Structural steel is the main contributor, to embodied carbon and should be a key focus of reduction in the modularization case, for example, by using only the required amount of transport steel reinforcement according to the shipment distance. McKinsey's report on decarbonization in steel and the Material Economics report on heavy industry decarbonization provide an overview of how steel production processes can be improved. For example, carbon-capturing from the blast furnace can result in a 50% reduction of emissions, carbon capture at the smelting reduction process can reduce 80% of the emissions. Biomass as a reductant, for example, substituted for charcoal in the blast furnace, can also create a more carbon-neutral option. Possible use of electric arc furnaces uses electricity to melt steel scrap with significantly fewer emissions, particularly where low carbon electricity is used.

Modularization for most parts of the project can be heavily utilized for LNG and petrochemicals plants. Modularization can reduce schedule delays and manpower at the site. However, decarbonization of modularization needs to be applied and heavily controlled at the yards. The benefits of modularization for sustainability depend on fabrication yards applying efficient production methods and utilizing clean energy. Engineering and design solutions should also optimize module size to reduce plot size and quantity of required concrete, structural steel, and piping materials. Proper assembly and fabrication processes, efficiency and gain in productivity in the fabrication yards should allow for additional carbon footprint benefits such as proper waste management at the site with investment in local recycling capabilities, opportunities for reuse of excess materials between projects, inventory, and material management controls to reduce reorders, damage, and material waste.

To reduce the logistics footprint, modularization strategies should consider: selecting carriers / sub-contractors / vendors using alternative fuels with lower carbon intensity and using innovative technologies with reduced CO_2 emissions such as electrical self-propelled modular transporter (SPMTs). Strategies can also include minimizing the distance between fabrication yard and site to reduce emissions produced resulting from transportation, increasing local content, reducing air freight, or reducing additional steel required for sea fastening (seaworthy transport). During planning, projects can utilize 3D model software and other technology to plan and simulate module logistics. During the Front-End Engineering Design (FEED) stage, AI implementation should be developed with the best optimized and sustainable logistics, procurement, and materials movements, also including the taxes, and all other obligations within each county. Spacing on ships must be fully utilized and all support piping and equipment can be shipped within.

Another strategy to support modularization is for EPCI companies to incorporate Advanced Work packaging (AWP) into their strategies and project execution as early as possible. Constructability and AWP help reduce waste and energy by promoting more efficient sequential work planning, increasing productivity, and facilitating quality design. Constructability encompasses a start-to-finish review of the construction required to build the job. Constructability reviews should specifically add carbon footprint and sustainability as an objective, in addition to safety, cost, schedule, and quality. Digital AWP helps manage change, which can reduce waste and improve efficiency, as well as reducing paper waste and trips between office and site. It also helps with planning in the work area, such as labor and equipment required for each scope of work at each location.

A key opportunity in the front-end to improve sustainability is designing out material waste to reduce embedded carbon, energy input, and waste-related emissions. This reduces waste from changes and allows any excess materials to be used in other sections of the facility or other facilities. It also reduces scheduling issues, as components can be exchanged based on priority. Examples of standardization include: facility components, column, and steel or beam size and spacing, design of standard, adjustable cable type and size, design and maintain a digital warehouse – that will be available for engineers during the design stage, and they can build it based on what we already have, rather than from scratch. The engineering department should use simulation software to analyze all possible material sizes and use for particular plants, and the data should be combined with risk-associated matrices within procurement and fabrication. This can significantly reduce the delays and logistics of materials.

The Front-End Engineering Design (FEED) stage can also consider inputs of recycled material, management of construction waste (including local regulation and infrastructure considerations), and opportunities to reduce the overall materials required for the project.

Oversupply of materials and unused materials can be challenging for fabrication yards and construction sites. The movement and management of materials can impact carbon footprint, particularly for waste and transportation. Digital tools can help manage materials more efficiently, for example through an automated worksite using radio-frequency identification (RFID), global positioning system (GPS), barcode, drones, augmented reality, and blockchain technologies.
Use of digital tools to manage materials can reduce oversupply of materials and using RFID to track material movements, provide real-time inventory (reduce re-ordering), and automate checkout processes, using drones to deliver materials or tools to site areas, augmented reality to assist personnel with goods in the inspection or pick-and-place tasks. Beyond digital tools, other simple opportunities can reduce waste in materials management. Intermodal warehouse systems (IWS) are steel pallets that enable vertical storage, saving space, and that can be more easily reused over wood pallets and be designed and packed by advanced work package. Optimized spacing can halve the space required for warehousing. Reduced warehousing space overall would save energy for lighting, heating/cooling, and other building emissions.

Those are examples of energy transition which mostly are combined with technology and new way of approaching oil and gas industry. First transformation is possible if digitalization and re-branding will take place to support the energy transition.

2nd Transformation: Digitalization

The oil and gas industry need to accelerate the use of newer technology to optimize the processes to support sustainability and reduction of carbon footprint. Transformation of the work with human-machine collaboration and identification of the new forms of collective intelligence to reduce man-hours and man presence at facilities will improve the processes. The oil and gas industry tend to master the processes and value experience over innovation due to safety and high risks. But while we go through difficult challenges on reducing carbon footprint and energy transition, EPCI and operators need to question every process. The only way you can do it is through fresh eyes that don't vision processes in the same way.

Application of digital tools in every cycle of a project is crucial to optimize the sustainable design and EPCI strategy. Artificial intelligence should be applied in the design phase, based on a series of algorithms and historical data from past projects. Artificial Intelligence, the use of programming languages like Python to estimate and predict the number of man-hours for upcoming projects, as well as the use of pattern recognition during the detailed engineering design, should be applied to recognize patterns of specific elements and reduce the number of mistakes and rework. The use of databases and collection of historical data with recognizing the delta of changes can predict the schedule development and give an overall picture and design of the upcoming projects. Standard dataset and estimation of carbon footprint, sustainability evaluation from techno-commercial standpoint, and tools to illustrate the impacts of design choices should be designed and included during the Front-End Engineering Design (FEED) stage to make sure that we are using the most optimized option to design facilities and developing the field.

Virtual and Augmented realities should be used in model reviews within 3D printers to develop the field and complex 3D model animation and software need to be used. AR/VR can allow more remote working, training, and inspection. 3D software needs to connect to the satellite and predict and design the plot plan based on real-time data to test EPCI strategies, understand potential material management issues, and optimize the construction, fabrication, and/or modularization processes. Collaboration with the tech giants is needed such as Microsoft and Google to develop low code applications that will fit the EPCI business requirements.

Digital libraries, communication platforms, and academies should be developed within the company to educate, connect and provide employees with necessary tools and information. Institute of knowledge needs to be transferred to the digital world and people can improve their skills in different time zones. Predictive application to build project teams based on historical data of employees and their rating should select the team for the project based on the set of skills required.

Use of drones to handle live progress reporting, precision surveying, health, safety, and environment (HSE) compliance, material laydown tracking, logistics planning, and incorporation of aerial site intelligence into 2D and 3D models. Drones technology is improving and can be used heavily offshore to transport necessary materials to platforms. Recent field tests have shown that drones can provide a fast, safe, and cost-effective solution for the inspection and maintenance of offshore oil and gas facilities. The technology is expected to be embraced by the growing offshore wind energy industry as well. Drones can provide safety inspection on vessels and at critical paths such as wind farms and power stations, oil spill detections and spill damage, gas leak detection, and work as facility security.

Digitalization can play a key role at the fabrication yards, particularly in reducing energy and waste. More meaningful reductions come with increased complexity or cost, for example, by advancing robotics in the fabrication process for more autonomous operations, robotic material handling, or robotic welding, cutting, or other manufacturing improvements. Similarly, autonomous vehicles could move materials and equipment, reducing waste from idling and underutilization. Sensors on equipment can acquire real-time data on location, fuel consumption, and utilization, helping drive efficiency, enabling predictive maintenance, and identifying high emissions activities or opportunities for the greatest reductions. Technology is the crucial element to reduce carbon footprint and optimize the process and prevent environmental tragedies.

Social media within the company needs to be strongly supported and used to collaborate and innovate. Movement of all meetings online and use of artificial intelligence to recognize the voice and taking notes during the meeting and converting voice to text should be used and distributed within the company. As well as the use of holographic meetings during the engineering design. For example, Microsoft is adding a "Spatial Rooms" tab to the Teams collaboration app. In Spatial Rooms, people can work on projects, customize the room, and come back to continue the work later. This will allow companies to move to the digital world and close the physical offices to continue working from home and eliminate transportation.

3rd Transformation: Cultural Innovation and Branding

Advances in communication technology and digitalization have fundamentally transformed the way humans collaborate and increased throughout all nations and industries as the Covid-19 pandemic arise. Remote work and virtual meetings are likely to continue, in a less intensely than at the pandemic's peak, but they are here to stay. While telecommuting has been possible for many years, remote work during the pandemic was supported by rapid deployment of new digital solutions, such as videoconferencing, document-sharing tools, social media platforms and expansion of cloud-based computing capacity. Nations across the world quickly designated essential workers who had to be on-site and told everyone else to stay home. The pandemic demonstrated that much more work could be done remotely than previously thought, including business sales calls, legal arbitration and trials, doctor visits, classroom learning, real estate tours, and even expert repairs of the world's most sophisticated machinery made with the help of virtual reality headsets. However, in the O&G industry, even with all this multiple digital platforms, 14% of permanent employees were laid off in 2020 and it is said that 70% of those jobs will not come back after the pandemic; leaving a bittersweet sentiment in and ongoing industry challenge in regards to retaining and attracting workforce. It is important to make the industry very attractive to users and employees. There are several digital transformational steps which companies need to get through to ensure employees are retained. Shifting the company's culture towards sustainability and digitalization, removal of the vertical hierarchical approach to achieve an innovative environment, and the setup of a communication platform for employees to create a horizontal approach. Companies will need to build a new workforce architecture to attract new talent. This includes, but is not restricted to, building career paths for new roles, such as agile coaches, data scientists, software engineering, back-end and front-end developers, emissions officers, or user-experience designers, to attract young talent.

Companies need to create a balance between experience and technology and innovation. Employees need to shift from experts-based to a cosmopolitan approach where employees have the opportunity to collaborate with different departments and learn and think wider than a project scope but for the entire shift of the industry. It is typical for EPCI companies to work from project to project and employees just master their main skills and rotate to the same position within several projects without noticing the shift of the industry and demand for new ideas and approaches.

That's why, a cultural shift needs to happen, and employees need to be encouraged to innovate and think outside of the box. Standardization is now in the past, and it doesn't improve efficiency when the market and industry are going through energy transition as well as digitalization transformation. Educational platforms need to be developed for employees to inspire and bring awareness on sustainability and digitalization. The creation of employee resource groups and platforms for people to learn and improve, socialize and collaborate is essential to shift the culture. Employee resource groups, awareness lectures, and how to change the way of doing things should be heavily incorporated to help current employees understand the market and adapt to this new emerging culture and ideology. Until organizations themselves undergo a fundamental change, including building a cohesion culture where employees have a sense of belonging and pride, their talent-acquisition landscape will not change. They need to feel that companies care about them and envision to keep them, help them to improve, and have career opportunities. Employees need to feel that companies are investing in them.

Although there is a minor trend toward hiring more graduates right now, the drop in the demand for specialists is a worrying sign as retirements cause accumulative loss of essential knowledge. Attaining the right balance between new and tenured workforce is crucial during those transformations. Oil and gas need a young generation; however, they have more tendency to worry about the future of oil and gas as well as world climate change. The total number of U.S. university graduates for technical courses like petroleum and geological engineering courses dropped by 15–21% from 2015 to 2019. This means there will be a great gap soon, when older generations retire. The younger generation is attracted to a career with information technologies, entrepreneurship and social media potential, such as blogging.

The industry needs to initiate change right from a pre-hiring and rising environmental concerns may be deterring millennials. Digitalizing operations or announcing net-zero long-term plans could boost interest among new graduates. Deploying branding campaigns focused on sustainability and investments in artificial intelligence and machine learning can attract more workforce. One solution is to take advantage of technology that most prospective candidates are already using – social media. There is a strong need to educate the public due to the massive amount of misinformation surrounding the oil and gas industry. Social media is a powerful tool that can start a conversation with those who are against or neutral to the oil and gas sector.

People have always cared more about what their peers had to say on relevant topics than news organizations, institutions, and governments. By providing a portal through which industry professionals can connect, share knowledge and engage one another in the discussion, LinkedIn has become a model for collaboration that others can and should emulate. But collaboration is the only aspect of the value social media brings to the industry. Industry demographics will change in terms of the younger generation coming on board. With this new blood, the use of social media will increase.

Some oil & gas companies are using social media for branding purposes (which is one aspect of marketing). Companies need to maintain a presence on Instagram, YouTube, Vimeo, Facebook, Linked In, and Twitter to showcase projects, initiatives, photos, update industry and followers on news and events, and highlight employee activities. This is a great way on showcasing work and expertise to get the invitation to bids. Many renewable companies and startups need to have the opportunity to collaborate with oil and gas giants.

For example, CEO of BP, Bernard Looney wants to engage directly with society, especially with younger people and those who disagree with BP. "I know a lot of people have views on oil and gas companies and our role in the energy transition," he says in one early post. "I would like to use this platform to talk openly about that and explain the role BP can play, as I believe we share the same concerns and hopes."

In the event of a crisis such as an oil spill or pipeline failure, social media can provide a channel through which information can be transmitted and a PR disaster averted or at least contained. Social media can be used directly to focus primarily on collaboration and recruiting, as well as event participation, education, branding, and crisis communication. As the industry becomes more familiar with social media – particularly as younger people enter the scene – I'm sure even more uses will be found.

It is also important to keep emphasizing the importance of increasing diversity within operations and data science. Tackling some of the greatest challenges of the Agenda for Sustainable Development - from improving health to combating climate change- will rely on harnessing all talent. That means getting more women working in these fields. Diversity in research expands the pool of talented researchers, bringing in fresh perspectives, talent, and creativity. Women can play a critical role in science and technology communities and that their participation should be strengthened. Women were rated as excelling in taking initiative, acting with resilience, practicing self-development, driving for results, and displaying high integrity and honesty. They were thought to be more effective in 84% of the competencies that we most frequently measure.

All those transformations need to be led by strong and innovative leadership and investment in the new technology and research. Energy Transition is a great way to adapt to new realis, however, the shift needs to be significant. Without innovative and brave leadership and major investments, traditional EPCI companies will become less relevant and get left behind.

Conclusion

Globalization and nations' interdependence in addressing key matters such as climate change has brought massive disruption to workforce management in the past twenty years with an exponential disruption seen after Covid-19 begun.

A drop in oil demand production, rates of return (ROR), increase in renewables solutions, and concerns of the young generation about climate change and environmental problems will continue to be key problems the O&G Industry faces in the upcoming decades.

The Oil and Gas Industry will continue to be a key player in the energy sector, however with such problems increasing, it is a perfect time for companies to relook at the investments and re-strategize for their future. Perhaps the biggest problem with the oil companies' pledges is that avoiding dangerous warming will require drastic action within the next decade. Nevertheless, the plans are largely silent about the short-term. Industry accounted for only about 0.5 percent of global investment in renewable energy and a little more than a third of the investment in carbon capture technology. The shift in direction in short term to account for digital transformation, more sustainable solutions and a cultural change are needed for the industry to continue leading this energy sector. Drastic changes can happen only when those three transformations are combined and followed by oil and gas companies. It is time for the industry to recognize this subtle shift in reality – the world is different. As such, the oil and gas industry must adapt to technology and artificial intelligence, sustainability, and social media innovations simultaneously because they are interdependent and help to accelerate the processes.

Acknowledgements

This report draws from McDermott's internal research on low carbon oil and gas facilities.

References

McLachlan, Duncan. "Net Zero Upstream Facilities: Can We Afford Not to?" *Ioconsulting*, 25 Feb. 2021, www.ioconsulting.com/post/net-zero-upstream-facilities-can-we-afford-not-to.

Barbosa, Filipe. *Oil and Gas after COVID-19: The Day of Reckoning or a New Age of Opportunity?*, McKinsey and Company, 1 May 2020, www.mckinsey.com/~/media/McKinsey/Industries/Oil%20and%20Gas/Our%20Insights/Oil%20and%20gas%20after%20COVID%2019%20The%20day%20of%20reckoning%20or%20a%20new%20age%20of%20opportunity/Oil-and-gas-after-COVID-19-v4.pdf.

Sustainable Signals: Individual Investor Interest Driven by Impact, Conviction, and Choice, Morgan Stanley, 2019, www.morganstanley.com/pub/content/dam/msdotcom/infographics/sustainable-investing/Sustainable_Signals_Individual_Investor_White_Paper_Final.pdf.

(l_collins), Leigh Collins. "'We Will Be Pioneering': Saudi Arabia Reveals 50% Renewables Goal by 2030, But Is That Realistic?" *Recharge*, 29 Jan. 2021, www.rechargenews.com/energy-transition/we-will-be-pioneering-saudi-arabia-reveals-50-renewables-goal-by-2030-but-is-that-realistic-/2-1-954094.

"Plan for Climate Change and Environmental Justice: Joe Biden." *Joe Biden for President: Official Campaign Website*, 29 Oct. 2020, joebiden.com/climate-plan/.

Kusnetz, Nicholas. "What Does Net Zero Emissions Mean for Big Oil? Not What You'd Think." *Inside Climate News*, 30 Nov. 2020, insideclimatenews.org/news/16072020/oil-gas-climate-pledges-bp-shell-exxon/.

Droguett, Barbara X. Rodriguez. *Embodied Carbon of Heating, Ventilation, Air Conditioning and Refrigerants (HVAC+R) Systems*. 2019, digital.lib.washington.edu/researchworks/bitstream/handle/1773/44736/Rodriguez_washington_0250E_20736.pdf?sequence=1&isAllowed=y.

"Batteries on Board: Offshore Vessels Setting the Course." *DNV GL*, www.dnvgl.com/expert-story/maritime-impact/Batteries-on-board-offshore-vessels-setting-the-course.html.

Clean Cities Alternative Fuel Price Report. 2020, afdc.energy.gov/files/u/publication/alternative_fuel_price_report_jan_2020.pdf.

"Reduced Emissions." *Neste Worldwide*, 17 Nov. 2020, www.neste.com/products/all-products/renewable-road-transport/reduced-emissions.

"Meet Bernard, BP's New Chief Executive Officer: News and Insights: Home." *Bp Global*, www.bp.com/en/global/corporate/news-and-insights/reimagining-energy/meet-bp-new-chief-executive-bernard-looney.html.

"The Future of Work in Oil, Gas, and Chemicals." *Deloitte Insights*, www2.deloitte.com/us/en/insights/industry/oil-and-gas/future-of-work-oil-and-gas-chemicals.html/#endnote-35.

McDonnell, Burns &. "White Paper: Advanced Work Packaging." *White Paper | Advanced Work Packaging*, info.burnsmcd.com/advanced-class-construction/advanced-work-packaging.

SEAKEEPING EVALUATION OF A TRI-SWACH BASED ON CFD CALCULATIONS AND MODEL TESTS

GONG XIANG[1,2,4], RAJU DATLA[3], XIANBO XIANG[1,4]

[1]SCHOOL OF NAVAL ARCHITECT AND OCEAN ENGINEERING, HUAZHONG UNIVERSITY OF SCIENCE AND TECHNOLOGY, WUHAN, 430074, CHINA
[2]THE STATE KEY LABORATORY OF FLUID POWER AND MECHATRONIC SYSTEMS, ZHEJIANG UNIVERSITY, HANGZHOU 310027, CHINA
[3]DEPARTMENT OF OCEAN ENGINEERING, STEVENS INSTITUTE OF TECHNOLOGY, HOBOKEN, NJ, USA
[4]HUBEI KEY LABORATORY OF NAVAL ARCHITECTURE AND OCEAN ENGINEERING HYDRODYNAMICS(HUST), WUHAN, CHINA

Proceedings of the 26th Offshore Symposium, February 17th 2021, Houston, Texas
Texas Section of the Society of Naval Architects and Marine Engineers

ABSTRACT

AEGIR is a time-domain seakeeping CFD code that uses an advanced, high-order boundary element method (BEM) to solve the three-dimensional potential-flow and has been developed for several years. In this paper, the latest version of AEGIR is used to predict the seakeeping of the Tri-SWACH with and without side hulls in headsea regular waves respectively. The primary objective was to evaluate its accuracy of predicting seakeeping performance of the Tri-SWACH under regular headsea waves in AEGIR. A series of simulated time domain heave and pitch responses for Tri-SWACH with and without side hulls via AEGIR have been compared with corresponding model tests conducted in the high speed towing tank in Davidson Lab, Stevens Institute of Technology. A good agreement in terms of heave and pitch responses between AEGIR numerical predictions and experimental data shows the seakeeping prediction capability of AEGIR for Tri-SWACH Preliminary Design. Also, the simulated seakeeping of a Tri-SWACH is compared with a Tri-SWACH center hull. It is found that the effects of side hulls will increase the transverse stability of the Tri-SWACH without causing additional significant effect on the seakeeping performance of the Tri-SWACH.

Keywords: seakeeping; Tri-SWACH, side hulls; model tests; CFD

INTRODUCTION

Conventionally, the first step in the early stage process of ship design is to find a proper mother hull such as a known "good" seakeeping hull found in the historical data and past practice. The second step is to improve the design of the mother hull according to the specific requirement. By slightly changing the geometry of the mother hull, numerous hull forms can be generated. After that, the analysis of their hydrodynamic performance such as resistance and seakeeping are very important due to their influence on powering, propulsion, main size of the ship. This approach has greatly relied on that the quality and the range of existing data were good enough for guide decision-making.

In modern times, numerous innovations in hull form technology have been introduced into shipbuilding industry

such as refinements to the monohull form such as bulbous bows, stern flaps, alternative propulsors such as waterjets, ducted props, and tunnel drives, the continuous creation of new hulls such as trimarans, slender high-speed catamarans, SWATH ships, etc. Because of the limits and restrictions on hull form options to the very limited number of previous hulls, these new hull forms or technology have no existing data for reference. The only course available to designers of novel hulls has been to resort to more complex boutique design tools peculiar to each type of hull. An efficient numerical tool capable of fast defining hull geometry and evaluating resistance and seakeeping properties of a variety of non-conventional hulls would greatly enhance the early stage designer's creativity and efficiency, the designer's ability to evolve effective ship concepts.
Initially, low fidelity two-dimensional (2D) methods are proposed for resistance and seakeeping simulations (Stern, Carrica, Kandasamy, Gorski, O'Dea, Hughes, Miller, Kring. Milewski, Hoffman, Smith, Cary, Taylor, Peterson, Harris, & Monaco. 2007). Later, more generalized high-fidelity codes, such as commercial codes, ANSYS Fluent, Star CCM+, open source codes, OpenFOAM, and inhouse codes CFDShip-Iowa have been all widely validated for predicting the seakeeping performance of floating vessels in Shen, Wan & Carrica, 2015, Poundra, Utamab, Hardiantoa, Suwasonoa, 2017, Begovic, Bertorelo, Mancini, Stern, Carrica, Kandasamy, 2006. However, as Stern in 2007 mentioned, AEGIR, as a hydrodynamic analysis tool and ship motion simulator based on three-dimensional, potential-flow wave-body interaction theory offers a more computationally efficient solver compared to viscous flow CFD codes. Both linear and nonlinear formulations for the wave flow and body motions have been implemented. The Rankine Boundary Element Method (BEM) approach, which entails the solution of a boundary integral formulation on the body and free surface boundaries for the governing Laplace equation. To evaluate the suitability and capability of AEGIR for use in an early stage design environment and its potential for various non-conventional hull forms, numerous researches have been done numerically and experimentally. A team consisting of a college intern of Naval Research Enterprise Intern Program found the correlation between AEGIR and model test results for calm water performance (bare hull resistance, heave and pitch) which was excellent for the Joint High Speed Sealift (JHSS) monohull configurations but was poor for other hullforms studied such as the Trimaran Hull with a wide transom. Datla and Kim, 2009 gave the overall conclusion that AEGIR may only perform well to predict calm water and headsea seakeeping performance for hullforms with small or no transoms. Malia, 2018 compares AEGIR and Gazebo—an open source, real-time robotics simulator, with an extension that integrates a model for the hydrodynamic wave forces into the rigid-body dynamics engine for modeling and simulating the seakeeping of a small unmanned vessel. The results suggest that the simplified models generate vessel motions of acceptable fidelity for the development of autonomy, but that for certain wave environments, the differences are significant. Based on AEGIR, Carter, Muk-Pavic and McDonald, 2013 managed to apply Artificial Neural Networks into resistance prediction for preliminary Trimaran Small Waterplane Area Center Hull (Tri-SWACH) design.
In this paper, the objective is to evaluate the capability of AEGIR into predicting the seakeeping performance of Tri-SWACH hull. Firstly, the seakeeping model tests for Tri-SWACH with and without side hulls are conducted under regular headsea waves with different wave length and amplitude. Correspondingly, heave and pitch performances are obtained in the time domain. At last, the simulated heave and pitch responses via AEGIR are compared with corresponding model tests.

METHODOLOGY

AEGIR is a high fidelity time-domain seakeeping code that uses an advanced, high-order boundary element method (BEM) to solve three-dimensional potential flow problems emphasizing those involving ocean free surface boundary conditions. Also AEGIR includes a fully nonlinear steady-state solver for sinkage, trim, and wave resistance predictions. Firstly, based on the work of Kring 1994, Maniar, 1995 and Danmeier, 1999 , Kring, Korsmeyer, Singer, and White, 1999 applied Rankine panel method into Aegir to solve the time domain seakeeping problems. Recently, more functions and modules have been built into it such as modelling steady flows around the hulls of sailing yachts and multi-hull vessels by Stern 2007, Kring, Milewski , and Fine, 2004, simulating the seakeeping of multi-body configurations in Peltzer, Rosenthal, Reppun, Kring, Milewski, and Connell 2008, performing multi-disciplinary dynamic simulations of high-speed monohull in Hall, Milewski, Coopersmith, Kring 2009 and predicting the hydrodynamics optimization of high-speed multi-hulls. Under the assumption that the fluid is incompressible and inviscid, according to the potential flow theory, AEGIR solves a boundary integral equation (BIE) with the boundary defined as the no-flux wall condition on the body surface, the combined kinematic and dynamic conditions at the free surface. The BIE shown in Eq. (1) is obtained by applying Green's Theorem to the Laplace equation.

$$2\pi\Phi_s(\vec{x}) = \iint \frac{\partial\Phi_s(\vec{x})}{\partial n} G(\vec{x}',\vec{x})\mathrm{d}x' - \iint \frac{\partial G(\vec{x}',\vec{x})}{\partial n} \Phi_s(\vec{x})\mathrm{d}x' \quad (1)$$

Where $G(\vec{x}',\vec{x})$ is the Rankine source potential, $G(\vec{x}',\vec{x}) = 1/|\vec{x}' - \vec{x}|$. $\Phi_s(\vec{x})$ is the steady velocity potential on the defined nonlinear free surface profile.

The combined dynamic and kinetic boundary conditions on the free surface are represented by

$$\frac{\partial\Phi_s(\vec{x})}{\partial z} = \frac{1}{g}\left[\left(-\vec{U}(\vec{x}) + \nabla\Phi_s\right)\cdot\nabla\left(-\vec{U}(\vec{x}) + \nabla\Phi_s\right)\cdot\nabla\Phi_s\right] + \frac{1}{2g}\left[\left(-\vec{U}(\vec{x}) + \nabla\Phi_s\right)\cdot\nabla(\nabla\Phi_s\cdot\nabla\Phi_s)\right] \quad (2)$$

The boundary conditions are imposed in a moving frame of reference, $\vec{U}(\vec{x})$, which can include mean forward, side-slip and steady rotational speeds

High-order discretization
The BIE and boundary conditions are discretized using a high-order method based on tensor product B Splines. The equations are integrated on the exact geometry defined by the CAD geometry input to AEGIR. Typically Bi-quadratic B-Splines are adopted to represent velocity potential distribution on the body surface and free surface. As described in Nakos 1990, as one important component, the use of the high order representation of the solution on the free surface realize the implementation of non-dissipative free surface conditions. As a result, more robust and accurate computations of the body generated waves over the entire computational domain can be realized. Also, a "geometry-independent" approach, uses discretization of the flow that is separate from the geometric description. In this approach, the expression for the wave flow is decoupled from the expression for the geometry.

Non-Uniform Rational B-Splines (NURBS) -based geometry capability
AEGIR has an integrated set of computational geometry algorithms for automating discretization of surfaces on the body and free surface as part of code execution and includes the functionality to directly read files in the native Rhinoceros format, providing the capability to quickly set up analyses of new concepts given a CAD representation. The integrated NURBS -based geometry capability is a key element to enhancing the utility of the AEGIR for concept design studies and hullform design optimization. The integral formulation is applied precisely on CAD geometry as developed by the designer.

Limits of application
One of the common problem with any potential flow method is dealing with the transom properly because of strong viscous effects in this region. In AEGIR, a detachment scheme conceptually similar to the Kutta condition is used for lifting flows. When a smooth detachment of the flow from the transom can be anticipated, namely the hull has no or a narrow transom, AEGIR is found dealing with transom condition with reasonable engineering accuracy at large Froude numbers (based on transom depth). However, when free surface conditions are used for wide transom, problem occurs because the wave elevation actually just match the mean free surface rather than the transom depth.

MODEL TESTS

Model description
Tri-SWACH is Trimaran-Small Water Plane Area Center Hull. The Tri-SWACH combines the concepts of a trimaran and Small Waterplane Area Twin Hull (SWATH) hull forms. Tri-SWACH hull forms have slender side hulls outboard a single small waterplane center hull. The side hulls provide stability while the center hull has a minimal water plane area to reduce wave excitation and hence motions and a very narrow transom as shown in Fig.1 to delay the flow

separation. Compared to a similar displacement monohull, the Tri-SWACH hull form has increased wetted area (and hence frictional resistance) however the very slender waterplane significantly reduces residuary resistance. The geometry details including both model and prototype are listed in table 1:

Fig.1 Local view at the transom of Tri-SWACH center hull

Table 1: Main Parameters of Tri-SWACH Hull

PARAMETER	MODEL	UNITS	PROTOTYPE	UNITS
λ	1		25	
Δ	13.200	kg	206155.917	kg
LCB	1.018	m (aft)	25.457	m (aft)
VCB	0.052	m	1.311	m
KML	2.025	m	50.640	m
KMT	0.150	m	3.75	m
LCF	1.093	m (aft)	27.317	m (aft)
LWL(main hull)	2.019	m	50.488	m
LWL(side hull)	0.551	m	13.78	m
Beam	0.060	m	1.494	m
Wetted Surface Area	0.778	m^2	486.215	m^2
Location of Out Riggers	Mid			

Fig. 2 shows the 1:25 model scale Tri-SWACH Center hull which is ballasted to a specified loading condition with 0.1m draft and 0-degree trim by placing ballast weights on weight pegs located in the model.

Fig.2 Tri-SWACH center hull set up in the towing tank

Headsea Regular Wave Seakeeping Test

The seakeeping model tests are conducted in 20°C fresh water in Davidson Laboratory's high speed towing tank, Tank 3 in Stevens Institute of Technology which is 95m long, 5m wide, and 2m deep. The towing carriage rides on a monorail situated above the centerline of the tank. The monorail is 45 inches above the water's surface. A steel cable driven by an electric motor at the far end of the tank drives the carriage. The wave maker is an articulated double-flap wet-back type unit. The upper and lower flaps are powered by hydraulic cylinders. The actuators are computer controlled and are capable of generating both regular and irregular waves. Firstly, the Tri-SWACH hull is towed under a set of wave conditions at the speed of V=1.335m/s, Fr=0.3 as listed in Table.2. During tests, two video cameras: Cam 1 and Cam 2 which are mounted at the front and back of the model respectively via carriage can clearly capture the scene of Tri-SWACH hull running against waves as shown in Fig.3 and Fig.4. The Tri-SWACH hull was free to heave and pitch but fixed in yaw, roll, surge and sway. The vertical motion of the tow-point was measured using a motion transducer attached to the free-to-heave apparatus. Pitch of the model mounting pad relative to the horizon was measured using an inclinometer mounted inside the model. Resistance was measured using a drag balance located directly above a pivot box (towing fixture with hinge permitting pitch motion). These heave, pitch and drag sensors can accurately measure the parameters in time and output the data through 3 channels to the core computer in an electronic signal way. Calibration for these sensors are completed before the model tests start. Similarly, the Tri-SWACH center hull is running in a series of regular headsea waves as shown in Table.3.

Table 2: Experimental matrices for Tri-SWACH

Run NO.	Vm(m/s)	T(s)	A(m)	L(m)
210	1.335	2.5	0.0145	8.70
217	1.335	2	0.0208	6.04
218	1.335	1.54	0.0107	3.72

Table 3: Experimental matrices for Tri-SWACH Center Hull

Run NO.	Vm(m/s)	T(s)	A(m)	L(m)
242	1.335	2.5	0.0305	8.74
248	1.335	2	0.0213	6.12
260	1.335	1.54	0.0109	3.70

Fig 3. Scene captured by Camera ahead of Tri-SWACH Hull during Seakeeping Test

Fig.4. Scene captured by Camera at the back of Tri-SWACH Hull during Seakeeping Test

REGULAR WAVE HEADSEA SEAKEEPING SIMULATIONS

Numerical model

AEGIR automatically builds the body and free surface grids based on a few user specified parameters related to domain size and panel density. The typical size of free surface domain is 3 ship lengths by 1 ship-length in Kring Milewski, Connell, Petersen, Kirschner, and Willemann, 2010. For monohulls, the software typically required three categories of patches: outer free surface, transom free surface (water surface behind ship's transom), and the hull. Fig 5 and Fig 6 show a typical extent of computational domain and numerical grid from a front and top view. The ability to directly import the CAD geometry from Rhino and the automated gridding process significantly reduced the workload for the user. After computation, AEGIR generates multiple files for data analysis.

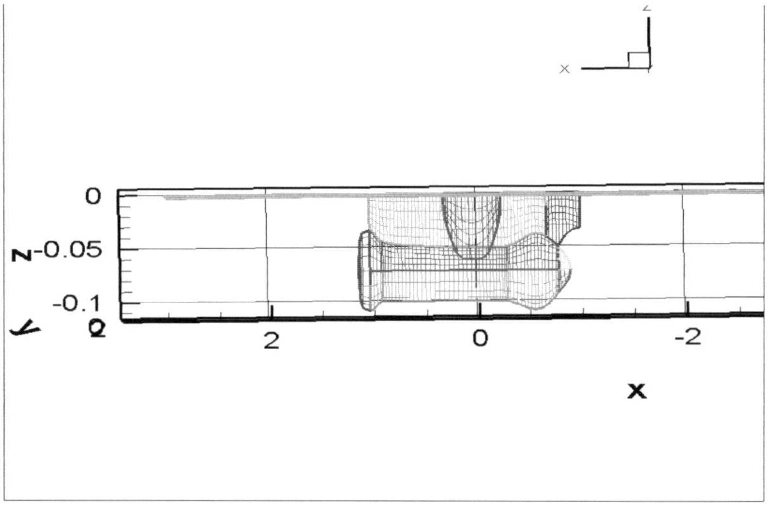

Fig.5 Frontal view of grids in Tecplot

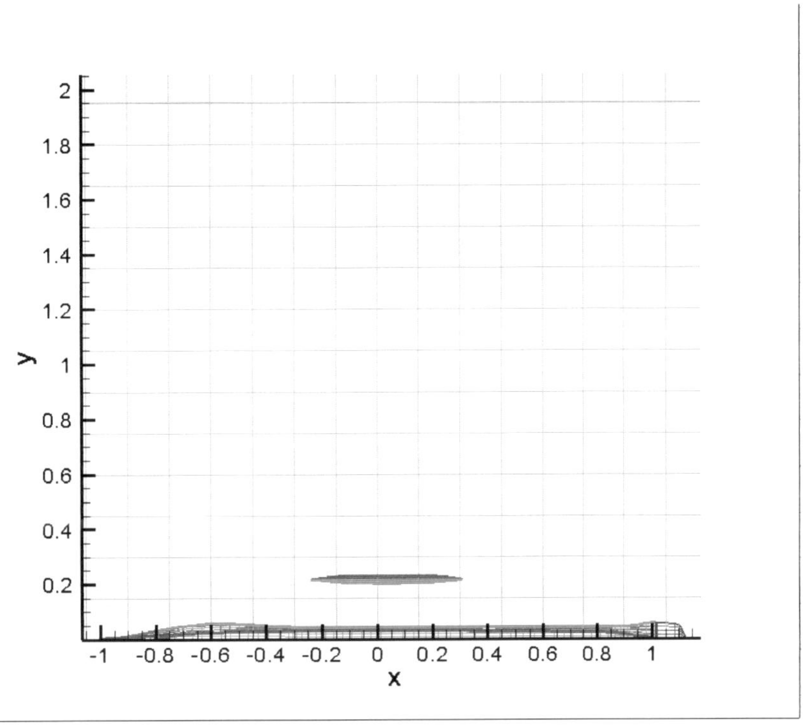

Fig.6 Top view of grids in Tecplot

Heave performance

The Tri-SWACH hull and the center hull are running with free pitch and heave motions under the regular headsea wave conditions corresponding to Table.2 and Table.3. It is noted that the pitch and heave responses presented in the following figures are all normalized by the wave amplitude. Fig.7(a), Fig.8(a), and Fig.9(a) present the comparison of heave responses for Tri-SWACH hull between AEGIR and model tests. Fig.7(b), Fig.8(b), and Fig.9(b) compare the heave responses for Tri-SWACH center hull predicted by AEGIR with model tests. It is found that for both Tri-SWACH hull and the center hull, the heave responses predicted by AEGIR match the data from model test very well although the discrepancy increases with the wave period decreases. Also, with the decrease of wave period, the heave motions becomes more frequent and irregular. The amplitude of the heave motions starts to increase. This may be because the natural frequency of heave motion is 6.85rad/s. With wave period decreasing from 2.5s to 1.54s, namely wave frequency increasing from 2.5 rad/s to 4.08 rad/s, the wave frequency becomes closer to the natural frequency, as a results, larger heave RAOs are observed in both aegir simulation and model tests. By comparing the heave responses for Tri-SWACH hull and center hull, the existence of side hulls seems cause the larger heave motion responses of Tri-SWACH hull.

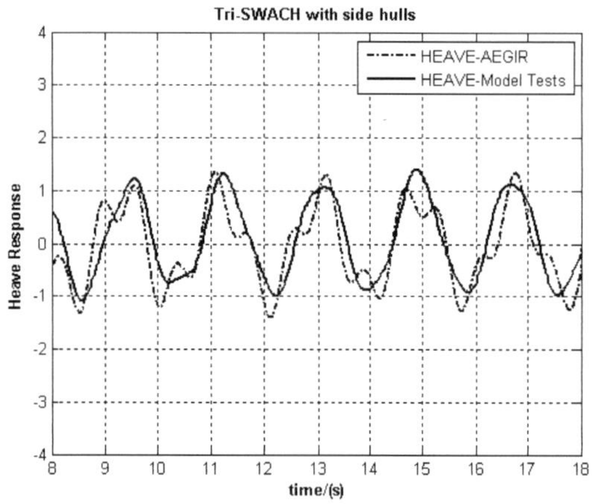

(a)

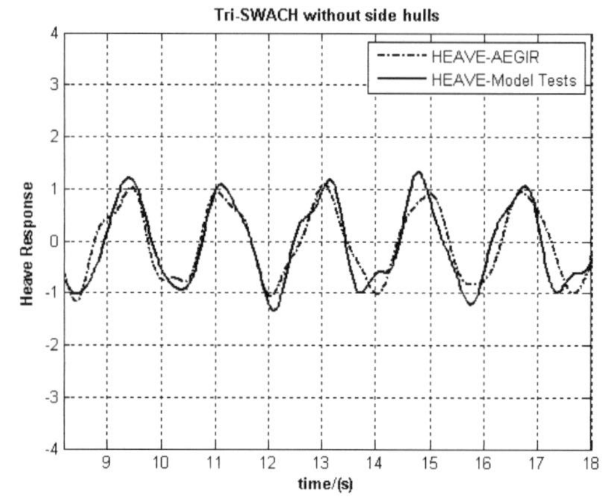

(b)

Fig. 7. Time domain record of heave responses at T=2.5s

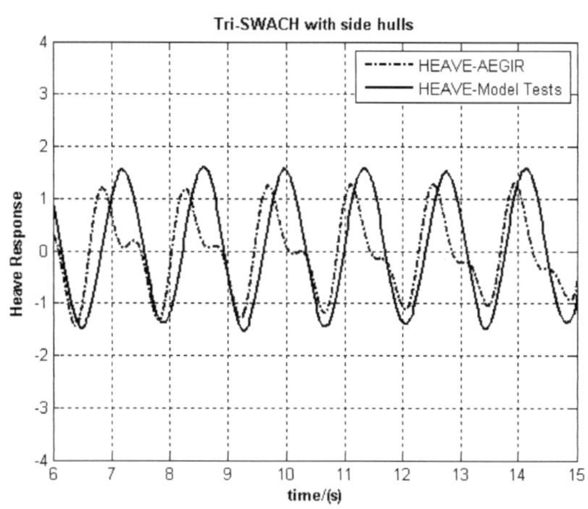

(a)

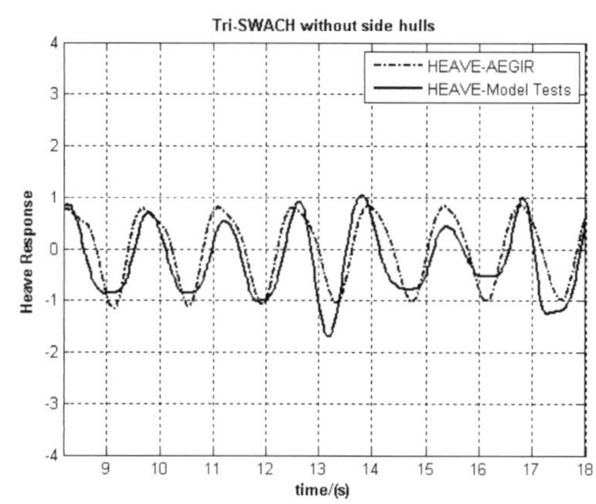

(b)

Fig. 8. Time domain record of heave responses at T=2s

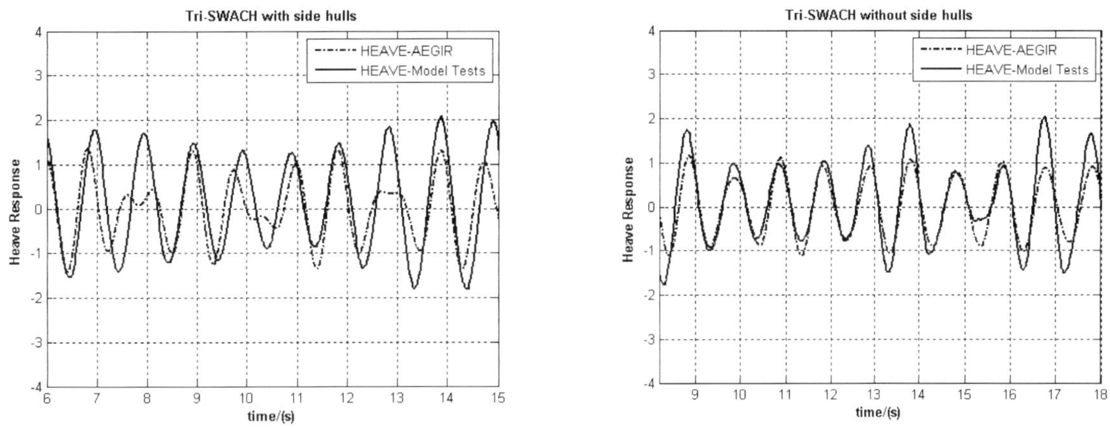

(a) (b)

Fig.9. Time domain record of heave responses at T=1.54s

Pitch performance

Fig.10(a), Fig.11(a), and Fig.12(a) compare the AEGIR predicted pitch responses of Tri-SWACH hull under waves with T=2.5s, 2.0s, and 1.54s to corresponding model tests according to experimental matrix in Table.2 and 3. Fig.10(b), Fig.11(b), and Fig.12(b) show the comparison of pitch responses for Tri-SWACH center hull. Due to the Tri-SWACH hull is so heavy compared with the required displacement that an unloader spring was used to give some lift for the Tri-SWACH hull to meet the required draft. But at the meantime, the spring will reduce the sensitivity of the model's pitch motions. The discrepancy increases with the decrease of wave period. Also, like heave motions, with the wave period decreases, the pitch motions become more irregular. The amplitude of the pitch motions starts to increase. This is also because the natural frequency of pitch motion is 7.04rad/s. With wave frequency increasing from 2.5 rad/s to 4.08 rad/s, the wave frequency becomes closer to the natural frequency. By comparing the pitch responses for Tri-SWACH hull and center hull, the existence of side hulls seems not to cause significant effect on the pitch motion responses of Tri-SWACH hull.

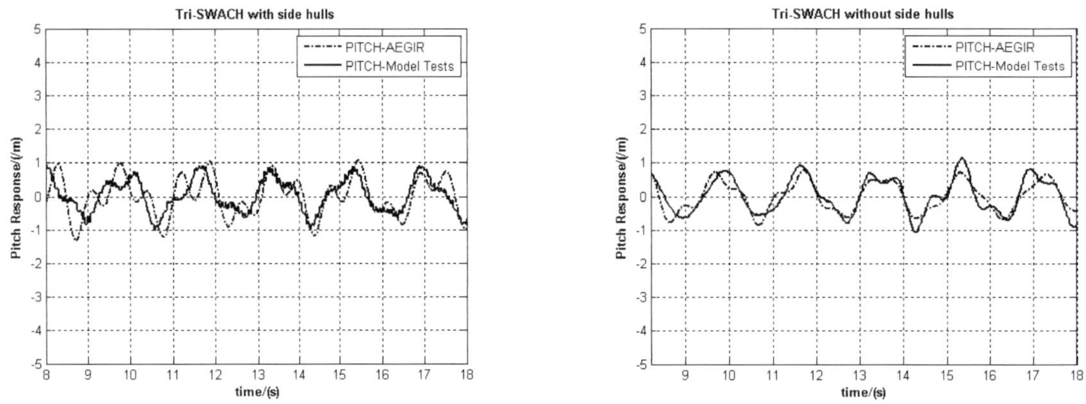

(a) (b)

Fig.10. Time domain record of pitch responses at T=2.5s

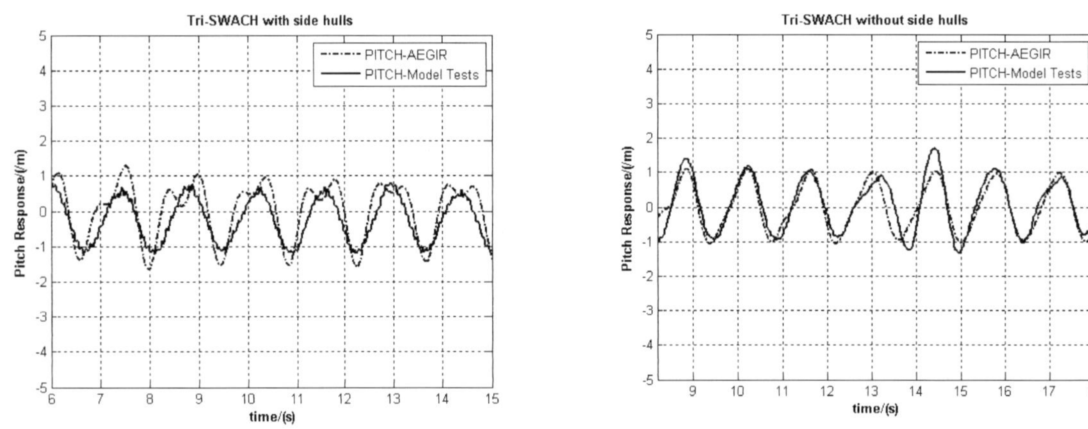

(a) (b)

Fig. 11. Time domain record of pitch responses at T=2.0s

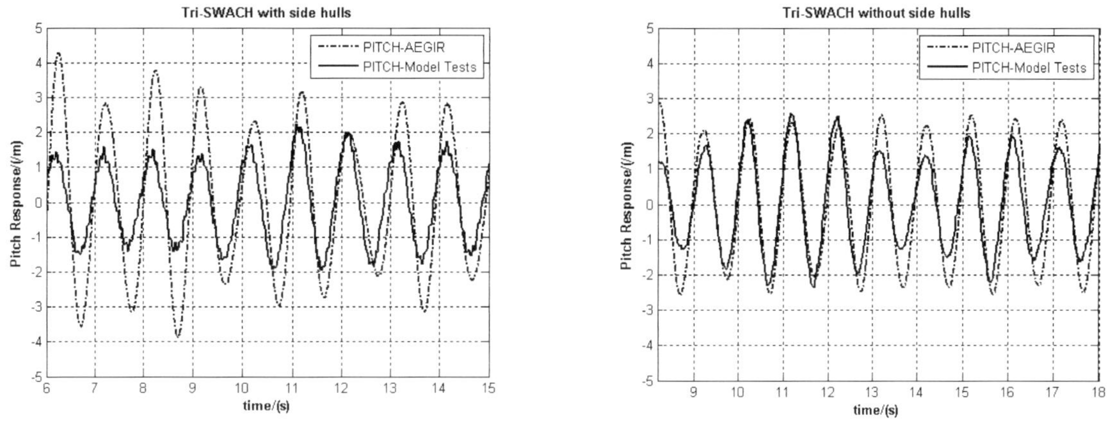

(a) (b)

Fig. 12. Time domain record of pitch responses at T=1.54s

CONCLUSIONS

This paper mainly focuses on the study of seakeeping performance of Tri-SWACH hull in head sea regular waves numerically using AEGIR and experimentally. The good agreement of predicted results with corresponding experimental data especially in the low frequency waves indicates the applicability and capability of AEGIR for seakeeping predictions in preliminary design. The heave and pitch motions of the Tri-SWACH hull and the center hull are both affected by the wave frequency. In the studied case, the side hulls are found not significantly affecting the seakeeping performance of the Tri-SWACH hull but in the future, the location of side hulls should be varied along longitudinal and transverse direction to study the effect on the seakeeping performance of the Tri-SWACH hull so that a more general conclusion can be drawn.

ACKNOWLEDGEMENTS

The authors wish to express our appreciation to Kelly Cooper for the generous support of this study. The authors also thank Justin Lorio for his assistance with the model test. The first author is partially funded by the Open Foundation of the State Key Laboratory of Fluid Power and Mechatronic Systems.

REFERENCES

Stern F, Carrica P, Kandasamy M, Gorski J, O'Dea J, Hughes M, Miller R, Kring D, Milewski W, Hoffman R., Smith S, Cary C, Taylor F, Peterson B, Harris D, Monaco C. Computational Hydrodynamic Tools for High-Speed Sealift: Phase II [R]. Final Report for Office of Naval Research grant N00014-06-1-0771 covering period 1 June 2006 to 30 September 2007

Shen Z R, Wan D C, Carrica P M. Dynamic overset grids in OpenFOAM with application to KCS self-propulsion and maneuvering [J]. Ocean Engineering, 2015(108):287-306

Poundra G A P, Utamab I K A P, Hardiantoa D, Suwasonoa B. Optimizing Trimaran Yacht Hull Configuration Based on Resistance and Seakeeping Criteria[J]. Procedia Engineering 2017(194):112-119,

Begovic E, Bertorelo C, Mancini S. Hydrodynamic performances of small size SWATH craft [J]. Brodogradnja, 2015, 66 (4):1-22

Stern, F., Carrica, P., Kandasamy, M., Gorski, J., O'Dea, J., Hughes, M., Miller, R.,Kring, D., Milewski, W., Hoffman, R., Steve Smith, Charles Cary, Philip Taylor Bill Peterson, Dan Harris, Curt Monaco [J]. Computational Hydrodynamic Tools for High-Speed Sealift. Transactions SNAME, 2006(114).

Datla, R, and Kim, H Y. Evaluation of a CFD Program AEGIRTM for Bare Hull Resistance and Seakeeping Prediction Capability [R]. Technical Report No. 2009/010. Naval Surface Warfare Center Carderock, West Bethesda, Maryland. 2009.

Malia J F.Modeling of A Dynamic Wave Environment for Unmanned Surface Vessel Control [D], Submitted in partial fulfillment of the requirements for the degree of master of science in mechanical engineering from the naval postgraduate school, December 2018.

Carter, A, Muk-Pavic E, and McDonald T. Resistance Prediction using Artificial Neural Networks for Preliminary Tri-SWACH Design [J]. Trans RINA, Intl J Maritime Engineering 2013(30)

Kring D C, Korsmeyer F T, Singer J, and White J K, Analyzing Mobile Offshore Bases using Accelerated Boundary-Element Methods[C]//VLFS'99 Conference, Honolulu, Hawaii, 1999.

Kring D C., Time Domain Ship Motions by a Three-Dimensional Rankine Panel Method [D], Ph.D. Thesis, MIT, Cambridge, MA, 1994.

Maniar H D, A three dimensional higher order panel method based on B-splines [D], Ph.D. Thesis, Massachusetts Institute of Technology, 1995.

Danmeier D.G., A Higher-Order Panel Method for Large-Amplitude Simulations of Bodies in Waves [D], Ph.D. Thesis, Massachusetts Institute of Technology, Cambridge, MA, 1999.

Kring D.C., Milewski W.M., and Fine N.E., Validation of a NURBS-based BEM for multihull ship seakeeping [C]// Proc. 25th Symposium on Naval Hydrodynamics, St. John"s, Newfoundland, Canada, 2004.

Peltzer T.J., Rosenthal B.J., Reppun W.K., Kring D.C., Milewski W.M., and Connell B., Multi-body Seakeeping Design Optimization [C]// Proc. 8th International Conf. on Hydrodynamics, Nantes, France, 2008.

Hall N., Milewski W.M., Coopersmith R.M., Kring D.C., "Analytical Modeling and Simulation of Water Based Aircraft Using a Time-Domain Seakeeping Code [C] // Proc. SAE 2009 Aero Tech Congress & Exhibition, Seattle, WA, 2009.

Nakos D.E., Ship Wave Patterns and Motions by a Three Dimensional Rankine Panel Method [D]. Ph.D. Thesis, Massachusetts Institute of Technology, Cambridge, MA, 1990.

Kring D, Milewski W, Connell B, Petersen B, Kirschner I, and Willemann S. AEGIR™ version 0.90 Time-Domain Seakeeping Program: User's Guide [R]. Oct,2010

AUTHOR INDEX

Society of Naval Architects and Marine Engineers (SNAME) Texas Section
99 Canal Center Plaza, Suite 310
Alexandria, VA 22314
USA

ISBN 978-1-7138-3136-5

9 781713 831365

CURRAN ASSOCIATES INC.
proceedings.com